PRESENTED TO

THE PEOPLE OF JACKSON COUNTY

IN MEMORY OF

ALFRED S.V. CARPENTER

1881-1974

The Environment

THE ENVIRONMENT
A History of the Idea

Paul Warde, Libby Robin,
and Sverker Sörlin

JOHNS HOPKINS UNIVERSITY PRESS   BALTIMORE

© 2018 Johns Hopkins University Press
All rights reserved. Published 2018
Printed in the United States of America on acid-free paper
9 8 7 6 5 4 3 2 1

Johns Hopkins University Press
2715 North Charles Street
Baltimore, Maryland 21218-4363
www.press.jhu.edu

Library of Congress Cataloging-in-Publication Data

Names: Warde, Paul, author. | Robin, Libby, 1956–, author. | Sörlin, Sverker,
    author.
Title: The environment : a history of the idea / Paul Warde, Libby Robin &
    Sverker Sörlin.
Description: Baltimore, Maryland : Johns Hopkins University Press, 2018. |
    Includes bibliographical references and index.
Identifiers: LCCN 2018007462 | ISBN 9781421426792 (hardcover) | ISBN
    9781421426808 (electronic) | ISBN 142142679X (hardcover) | ISBN
    1421426803 (electronic)
Subjects: LCSH: Environmental sciences—Philosophy. | Human ecology. |
    Nature—Effect of human beings on. | BISAC: SCIENCE / History. | SCIENCE /
    Life Sciences / Ecology.
Classification: LCC GE40 .W37 2018 | DDC 304.2—dc23
LC record available at https://lccn.loc.gov/2018007462

A catalog record for this book is available from the British Library.

*Special discounts are available for bulk purchases of this book. For more information,
please contact Special Sales at 410-516-6936 or specialsales@press.jhu.edu.*

Johns Hopkins University Press uses environmentally friendly book materials,
including recycled text paper that is composed of 30 percent post-consumer
waste.

Contents

Acknowledgments

This book was first conceived at the inaugural World Congress of Environmental History in Copenhagen in 2009. We three authors from the United Kingdom, Australia, and Sweden found ourselves talking about the idea of "the environment" and its history. The World Congress revealed that *environmental history* did not mean the same thing in different places. The environmental movement was central to some historical discussions; in others it was nearly absent. In the United States, environmental historians were *historians* whose work mapped the emergence of a social movement (called environmentalism since the 1970s, with various precursors). Elsewhere, environmental history included practitioners from a range of different disciplines: the environmentalism of the recent past was a subject for political science, while geographers and ecologists wrote about physical change over time, and historians wrote about the cultural history of these changes, and sometimes about Big History—the stories of the planet from the big bang to the present.

The scales of environmental history were variable. Yet the idea of "the environment" itself has its own history, which emerged in most Western places in the postwar years, strengthening in the second half of the twentieth century. Led by public scientists, the environment was tied closely to "managing" natural resources and the land. The environment was a global idea, fostered by a universal science and growing in importance at the same time as globalization itself. The environment framed the idea of the global (especially the planetary) as much as the new globalism shaped the interdisciplinary sciences that came to call themselves environmental starting in the early 1960s.

We decided to write a book about this big idea, which by 2009 was already changing shape again with the proposal of a new epoch in Earth's history, the Anthropocene. This was part of a project called "Expertise for the Future," where we were especially interested in how discussions about the environment always seemed drawn toward its future fate, and how particular people had gained authority in those conversations. Our ideas were shaped and tested across a series of workshops held at the University of East Anglia in the United Kingdom, Harvard University, the Australian National University, and the Royal Institute of Technology in Stockholm. We would like to particularly thank all of the participants in those events for their papers, questions, reflections, and discussions that extended well beyond the events themselves. We would also like to thank all of those who helped us organize the meetings.

These represent only a part of our accumulated intellectual debts over a period of nine years. They include, directly and indirectly, the authors of the many books mentioned in the bibliographic essay and notes. We benefited greatly from the many helpful comments on interim presentations from this book at conferences of the American Society of Environmental History, the European Society for Environmental History, and the second World Congress of Environmental History (Guimarães, Portugal, 2014), and in numerous other workshops and meetings. Financial and logistical support has come from the Centre for Environmental History, Australian National University; Center for History and Economics (Cambridge and Harvard); Division of the History of Science, Technology and Environment, Royal Institute of Technology (KTH), Stockholm, including its Environmental Humanities Laboratory; Fenner School of Environment and Society, Australian National University; Harvard University Center for the Environment; School of Social Science at the Institute for Advanced Study, Princeton, New Jersey; Peter Wall Institute for Advanced Studies, University of British Columbia; the IHOPE program (now at Uppsala University); Leverhulme Trust, UK; Rachel Carson Center for Environment and Society, Ludwig-Maximilians-University, Munich; Stockholm Resilience Center at Stockholm University; the

University of East Anglia; Riksbankens Jubileumsfond (Stockholm); and Formas—The Swedish Research Council for Environment, Agricultural Sciences, and Spatial Planning. At every step of the way, we have enjoyed the intellectual stimulation and conviviality of the Ely Institute.

All told, our (friendly!) critical commentators over the years have been too many to list here. The few that nonetheless deserve to be mentioned by name are those who used some of their precious time to serve as chairs and commentators at conference panels, or read full chapters for us at their various stages of completion (including when they weren't very complete at all), or commented on the structure of the project as a whole. These are Alison Bashford, Saul Cunningham, Tom Griffiths, Sabine Höhler, Sheila Jasanoff, Susanna Lidström, Gregg Mitman, Ed Russell, Anna Svensson, Jo Warde, Nina Wormbs, and Graeme Wynn.

The Environment

Prologue

The environment is all around us. This book asks the question, Where did it come from?

The environment is under threat as never before. Is it possible, we ask, for the economy to grow without the environment being destroyed? Will our lifestyles end up impoverishing the planet for our children and grandchildren?

Yet if we look back, within the lifetimes of many people alive today, such questions would have made no sense. This was not because we were having no impact on nature, nor because we were unaware of the fact of that impact. What we lacked was an idea: a way of imagining the web of interconnection and consequence of which the natural world is made. Without this, we also lacked a way to describe the scale and scope of human impact upon that world. This idea—a planet-changing idea, because it made the planet visible in a wholly new way—was "the environment."

This is a book about how the idea of the environment came to be and its consequences. We begin this story in 1948, in the ferment of reconstruction and recrimination, in the hope of new global institutions and the fear of humans' capacity for almost limitless destruction. It was at this moment that a new idea and a new narrative about the planet-wide impact of people's behavior emerged. "The environment" provided a concept that linked changes close to home to worldwide pressures. At a time of a new "world-mindedness," the environment became one of the concerns that nations shared—it was important for the raw resources that could create new peaceful societies. The conservation, restoration, and enhancement of the environment were part of international postwar reconstruction, and the environment shaped how many of the new global institutions developed.

In creating an object of imagination and measurement, it allowed a new kind of question about nature and human responsibilities to be asked. Is the world sick? Is the environment getting worse or better? But the environment changes over time, and so do the ideas about what it is. This depends on who speaks for the environment. Which people are concerned about it? How do people think about it?

The seven decades that have followed have seen vastly changed ideas about the environment. It has become much more complex. The environment is not just about the land but also the sea and the sky. The great oceans of the world and the Earth's atmosphere are important to life, but they are not the jurisdiction of single nations. The global environment provides habitats for plants and animals, including humans. The rise of megacities has changed the way humans live in the world. In the 1940s, most people lived close to where their food was grown, but now a majority live in cities, often very big cities in rapidly developing places like China and India.

The environment has gone from being the background to the (human) world to being an idea shaped by planetary consciousness. The Earth itself has become a "person," an agent of history. People talk differently about the environment because of this. In 1948 the environment emerged in a human historical context, and its institutions were about management and regulation. In 2018, we have so much more information and data from science, and it is freely available in colorful formats through the internet.

One picture from 1968 stands out as a change-maker: the view of the Earth from space, the blue planet—whole and exceptional in the solar system in being suitable for life. Earth is sometimes called the Goldilocks Planet—not too hot, not too cold, just right. Its exceptionalism calls for a new sort of care on a planetary scale. That people can see the whole planet in a single image has moved the concept of the environment from "international" to "global" and now "planetary." The context for life itself has become both bigger—and smaller. There is only one Earth.

What prompts change in environmental concerns at different times? Sometimes it is a historical moment, a disaster. For example, the accident on April 26, 1986, at Chernobyl nuclear power station in Ukraine heightened concern about how nuclear particles travel on the wind, how they go into the soil and affect food, what they mean for the future health of children growing up, and how the plants and animals that remain have responded to the traces of nuclear explosions in the environment. Each of these questions engaged different experts, people who measured the directions of the prevailing wind, how the nuclear particles were carried in the atmosphere, and how long it took for them to be absorbed. Agricultural and soil experts measured what might be safe to eat, and health scientists tried to improve the quality of future life for people exposed to the catastrophe. In the decades since, ecologists have measured how wild nature has responded and evolved in an area where people have been excluded. A "new nature" has grown up in this place of catastrophe: it will have a different future because of this moment. For many people ecologists have been the key experts of the environment. Indeed, in some places, people call the environment "the ecology."

As we trace the environment from 1948 onward in history and into the future, we find different experts are important at different times. Catastrophes have sometimes created the moment for a new expertise, but they are not always emergencies or "instant" disasters like Chernobyl. More often it is "slow catastrophes" that change the environment, working over longer time periods. Droughts and famines, the acidification of the oceans from agricultural chemicals, lead poisoning in children playing in areas with lead paint are all disasters, but they evolve more slowly than the human eye can see. Sometimes we know about the catastrophe only because of technical experts and specialist technologies. The thinning of the ozone layer in the atmosphere, caused by refrigerants called chlorofluorocarbons, is a good example of how expertise can help manage the environment: the ozone layer is "healing" because atmospheric chemists identified what was causing the hole and industry quickly changed the way refrigerators worked, using a different chemical that did not naturally combine with ozone. Generally, however, slow catastrophes are much harder to manage than crises and emergencies. There is more negotiation with more parties over longer times—and we often need many different types of expertise.

In this book, we have started with the postwar moment, when the world needed more food and building materials. Natural resource managers were important, and so were the new international negotiations that included concerns about the environment. Experts gathered around resources, conservation, and climate (see chapters 3, 4, and 5, respectively). The overall story reveals that the environment demanded interdisciplinary and multifaceted knowledge-making and understanding on many scales, as does managing and planning for the future. Yet each area of knowledge that fed into the new idea of the environment had its own history and set of techniques that in turn shaped the new understanding. We introduce them through important moments when they made a distinct contribution to the idea of environment, and we trace the history and controversies that put them in a position to make that contribution. More and more, the story we present is about the rise and rise of "integrated expertise," of Earth system science—which treats the Earth as a dynamic system, always in flux—and of Big Science, which joins up many different experts, and of the tools of modeling and measurement enhanced by the digital revolution, which accelerates across the same time span. The environment is about people, too, and how they respond to its changes and challenges, and it is not just experts who care about its future. Questions of justice for humans and for the environment itself are moral and complex.

So this is not so much a book about what the environment is as much as it is about what humans have wanted to make of it. It is about the imagination, the history, and the creativity of the experts, as well as their technical and diplomatic skills, and how well ordinary people have understood these and contributed to framing environmental concerns. Expertise has moved popular opinion, business practices, and understandings of the world, but only so far. Expertise has limits in the political world. We all share the environment and have a say in what is possible. Thus, the environment is a key concept: it drives conversations about what it means to be human in the world on many scales. These conversations include what sorts of responsibilities humans take for the results of their actions in the past, present, and future.

Road to Survival

A Fable for Tomorrow

Ask many people old enough to remember the first album of the Beatles and news of the blastoff of Yuri Gagarin what sparked their interest in the environment, and they will answer: Rachel Carson's book *Silent Spring*. "There once was a town in the heart of America," it began. America was perhaps at the peak of its economic might and, for much of the population, its home comforts. A vision of a fabled land, as sweet and harmonious as when first laid out by the doughty pioneers, leapt into the minds of millions as they began to read, perhaps already relaxing into a wistful reverie. But within a page, things began to change in the heartland. People fell sick. Doctors were mystified. Children out playing collapsed and died. And where were the birds? The skies were empty. The skies were silent. Was it witchcraft? Was it some monstrous enemy?

It was not. The creeping death was the product of science. It had been brought about by products designed to keep people healthy and the soil productive. They were made by the chemical industry, but it was ordinary Americans who picked up the products at the store and polluted their land, their wildlife, and their bodies. "The people had done it themselves."

First serialized in the *New Yorker* in June 1962, *Silent Spring* saw a vicious reaction from the chemical industry, which in the end only consolidated Carson's fame, with invitations for TV appearances, to the White House, and to testify before government committees that would affirm her claims. Her work galvanized activism and public policy in many countries. Carson herself would die within two years, at the young age of 56. But her legacy embraced and spanned the planet. Today, when the director of the United States Environmental Protection Agency steps forward to make any major pronouncement, he or she does so in the Rachel Carson Green Room.

Silent Spring famously described "the contamination of man's total environment with . . . substances of incredible potential for harm." In "the environment," a word she still felt the need to qualify, Carson found a term that could encompass the astonishing pervasiveness of chemical pollution that she catalogued.[1] She was trained in ecology, "a science teach[ing] us that we have to understand the interaction of all living things in the environment in which we live." This was her third book; she published two popular works on marine science in the postwar years and before that worked as a research scientist. Yet before alighting upon that so memorable title *Silent Spring*, she had considered using "The Control of Nature" or "Man against the Earth."[2] Carson considered the specific issues of pesticides to have much wider resonance, which was emblematic of humanity's more general relationship with nature.

At the same time, it seems as if in the years around 1960 she was still searching for a term that could highlight this relationship for a broader audience. "The environment" perhaps did not yet seem popular enough for a book title.[3] In fact, neither of her early suggested titles—using "control" or "against"—were very good descriptions of what she wrote about, as humans were clearly *not* controlling nature in the way that they might desire, while the risks she described were as much to people as to the Earth. "The environment" would come to signify dangers within us, in the forms of toxins, as much as outside, what we were doing to the planet, and might rebound upon us. Such was Carson's impact that, for many, her work later seemed almost a foundational moment for the term.

This book is a history of that idea of "the environment." This was not a new word or a new idea in 1962 by any means. But Carson's treatment of the idea does exemplify an enormous shift that had taken place in the years before she wrote. For many decades, *environment* had been a word used to describe the *context* or background to the real subject of the story, whatever that might be: a study of a species, a writer, a society, a race. It was shorthand for the set of unique surrounding circumstances, which might prove to be overpowering, serendipitous, exculpating, or promoting of adaptation or balance.[4] Of course, this is still a meaning we understand in the English language today. Things can be explained by "their environment." But that environment is not the real subject of interest—it is not really a thing at all. Carson, in contrast, wrote about "the environment," a thing with its own essence that itself became vulnerable, a victim of circumstances: as Carson put it, a fragile "web of life"[5] subject to contamination and assault, its "integrity" subject to "disturbance," to becoming "corrupt" and being "engulf[ed]."[6] There had been a shift from a world where "man" was "moulded by the environment" to him being able to "alter the nature of his world."[7]

Yet this transformation in meaning, and a whole environmental revolution, was not a product of the sixties. It was not a revelation bursting from a generation discovering protest, space travel, and sexual intercourse for the first time (to rehearse some other myths). Nor did the idea of the environment under threat detonate unexpectedly in the assured affluence and confidence of a land halfway—as we now know—through the term of its charismatic and zestful young president, John F. Kennedy. The environment was an idea whose moment came earlier, whose story was first written in an earlier time. In keeping with the foreboding and uncertainty associated with it, we must return to a world trying to put itself back together after a shattering war.

1948

January 1, 1948, a Thursday, was mild, wet, and windy in London, a frosty and sunny day in Berlin, and a rainy day with thunderstorms in Washington. In Moscow, it was full winter with a stiff ten degrees below zero. Six months later Berlin would be the hot center of a Cold War between these cities as the blockade and airlift, perhaps the most famous event of that year, began. The first of January was also the day when Britain nationalized its railways, when the first color newsreel was filmed in Pasadena, California, and when the General Agreement on Tariffs and Trade (GATT) became effective, signaling the ambition to build a world of free trade and openness—at least among the twenty-three signatory nations. An old order was dissolving; Sri Lanka and Malaysia expected their independence within weeks. Before the month was out, in the chaos and bloodshed of postpartition India, Mahatma Gandhi would be assassinated.

This was a world in the shadow of past and future wars. In vanquished Germany, starvation was common; looting and despair were still everywhere. The bitter and prolonged winter of 1947 had created fuel and food scarcities across the European continent. For some, the war had not just shown humans at their worst; it had shown them as they really are. In October, the Soviet Union presented a proposal to the United Nations to ban nuclear weapons. The proposal was rejected, and in late August the following year the Soviet Union detonated its own atomic bomb. Yet the natural world still offered some respite. George Orwell, amid the bomb sites and ration cards of postwar London, wrote how "the atom bombs are piling up in the factories, the police are prowling through the cities, the lies are streaming from loudspeakers, but the earth is still going around the sun, and neither the dictators nor the bureaucrats, deeply as they disapprove of the process, are able to prevent it." In 1948 he retreated to the solace of a cottage on the Scottish island of Jura, writing a book with the year's numbers in reverse in the title, *Nineteen Eighty-Four*.

Yet all was not well with nature. Shadows of another kind loomed ahead. They were modestly signposted in a book that appeared first in April in the United States and in June in England: *Road to Survival* by William Vogt.[8] This road was not the path of escape from the aggressions of the Axis powers or the pressures of metropolitan life. It was a different, daunting trail, a way to be chosen or not by everyone on the planet because everyone would have to walk it. "By excessive breeding and abuse of the land mankind has backed itself into an ecological trap," Vogt thundered. The present state of the world portended an existential threat, a trajectory much more all-encompassing than that offered by atomic weapons (as yet owned only by the United States). To write about the changes occurring to nature in 1948 was already to produce "A History of Our Future," as the final chapter of Vogt's book was named, an experience totally new to humanity.

Road to Survival presented a new history of the planet precisely because it was about the planet as a whole, and all the varied peoples that lived on it. It was a story of global interconnections, unmistakable since the war and its aftermath, but a story of ecological rather than political and military destruction. "An eroding hillside in Mexico or Yugoslavia," Vogt wrote, "affects the living standard and probability of survival of the American people. . . . We form an earth-company, and the lot of the Indiana farmer can no longer be isolated from that of the Bantu."[9] This was a new narrative about our planet; not of dreams of wealth, or ideological rifts, but about the very Earth that ailed beneath our feet. As Vogt declared in the preface to the English edition, "The world is sick."

Vogt presented the curve of the world's population, already rising steeply for some fifty years, and contrasted it with a graph of natural resources. These included topsoil, forests, water, grasslands, and "the biophysical web that holds them together." This curve of resources had been decreasing since industrialization, but now "it is plunging downward like a rapid." For Vogt, these two curves encapsulated the essence of his message and the fate of the world: they had crossed and were now drawing apart. If that gap could not be closed, "we may as well give up all hope of continuing civilized life. . . . Like Gadarene swine, we shall rush down a war-torn slope to a barbarian existence in the blackened rubble."[10]

Vogt did not come from nowhere. Trained in ecology at St. Stephen's (now Bard) College on the Hudson River in New York State, he was a dedicated ornithologist and a lifelong nature protectionist. He brought a scientific background to his cause, but also the practical skills of a negotiator. Yet like the later *Silent Spring*, *Road to Survival* was no scientific textbook: it was a book of passion and of outcry. It was written to *move* readers, not merely tell them things. The world of *Road to Survival* was one suffering the disease of overpopulation. Its symptoms were degraded soils, resource depletion, food scarcity, starvation, famine, and disease, all of which had accelerated during the recently ended war. The world in 1948 was being over-whelmed by modernity, by a humanity that had failed to set limits to what the human enterprise could achieve. Vogt, as much as any American on the threshold of the Cold War, was strongly anti-Soviet. But he was also deeply distrustful of capitalism.

Road to Survival quickly became an international bestseller, translated into nine languages and popularized by *Reader's Digest*. The book reached an estimated thirty million readers. Vogt became a well-known voice for population control, was appointed national director of the Planned Parenthood Federation of America from 1951 to 1962, and served as a scientific representative to the United Nations. The book was his rallying cry.

Vogt's writing was also novel in his use of the term *the environment*. In his thinking, *environment* ceased to mean context, just the local surroundings of an individual organism. Rather, in his mind, the environment became a global object. Indeed, it would come to embody the global in our minds, especially after that image of our planet, a shimmering blue-green orb in the darkness, became fixed by pictures taken by from the Apollo missions. This global object had a fundamental unity, and a single destiny. Would that spell doom or survival? Vogt wrote, "we live in one world in an ecological—an environmental—sense."[11] In this there was also a tacit priority: "environmental" concerns were the top value, the preeminent issue. Population growth was not so much a problem for the populations themselves, or for the nations and their well-being, but it had become a problem for the planet. High birth rates were not an issue for the health of mothers, fathers, or children but rather for the wealth and health of bogs, jungles, forests, and rivers.

The Accidental Revolution of "the Environment"

Road to Survival appeared at the beginning of a revolution in thinking. Part of the change was to make the insights of an ecologist political. Vogt's work privileged the expertise of the ecologist, combining his scientific training with views formed by wide travels and rich experience from fieldwork. He had spent several years in Latin America as chief of the Conservation Section of the Pan American Union, surveying population and resources. He cited examples from El Salvador, Costa Rica, Mexico, Venezuela, and Peru. Part of the power of his argument was to discover the trajectory for a whole planet everywhere, in places that people knew or could imagine. He connected, as we would later say, the local and the global. This was a notion that became embedded in the idea of "the environment" as he used it. There was an environment outside your door, and it was the same environment as the one outside any door on Earth, and what happened to each place had ramifications for all the others. The idea itself *scaled* up and down, it was inside and out, local and global. This was to become one of the key properties of environmental research in the second half of the twentieth century: it worked on many scales. The environment could appear on any level from the life-world of the microscopic organism to the entire world of humans, the Earth, and its atmosphere.

Vogt had corresponded with Aldo Leopold, another ecologist and the doyen of American conservation by the 1940s. *Road to Survival* was emblematic of a crucial generational change in thinking about conservation as a predominately localized issue to one where a planetary environment gathered together all the strings of all the environments that existed, and in turn made each of those—every mountain, wetland, forest, pond, bay, city—a subset of that whole. This thinking was prefigured in the concerns of Leopold, who had wondered if the war had been caused by the same ecological forces that led to localized extinctions of animals. Leopold died fighting a wildfire in the very month *Road to Survival* appeared, and his classic *Sand County Almanac* was published posthumously the following year. He wrote of a "land ethic," of the need to extend obligation and responsibility beyond the human realm to the environment. Conservation was an ethical issue; perhaps *the* ethical issue. Leopold declared: "There is as yet no ethic dealing with man's relation to land and to the animals and plants which grow upon it. Land . . . is still property. The land-relation is still strictly economic, entailing privileges but not obligations."[12] Vogt also echoed his fellow American ecologist Paul B. Sears, whose focus was soil degradation and "desertification" in the light of the midwestern Dust Bowl of the 1930s.

Ecology was a type of expertise that the world needed, although, Vogt added, it must be assisted by the social sciences, "the radar that can avert disastrous crashes."[13] This new thinking found voice in 1948, but it was also the product of developments in science and thinking during the interwar years. It had a polymathic style. *Road to Survival* evoked animal ecology, mathematical population models, and studies of soil degradation and bound them all together with prophecy. All the tools and wisdom of many disciplines would need to be brought to bear if we were to salvage the future fate of the world.

In that same year, Fairfield Osborn published the slim volume *Our Plundered Planet*. "The impulse to write this book," he noted in the introduction, "came towards the end of the Second World War. It seemed to me, during those days, that mankind was involved in *two* major conflicts. . . . This other war is man's conflict with nature."[14] This was not a war of conquest but of mutually assured destruction. Osborn was trained as a scientist with a degree from Princeton and an advanced degree in biology from Cambridge University. Son of renowned paleontologist (and racist) Henry Fairfield Osborn, the younger Fairfield worked as a businessman and had taken a strong interest in eugenics before turning his interest to nature conservation issues and especially the management of New York's zoo. From 1948 to 1961, he was the first president of the Conservation Foundation, an organization he founded with several like-minded colleagues to raise awareness about ecological problems. As with Vogt, his book won him fame and influential posts, including being asked to serve as a governmental advisor.

Of course, the title of *Our Plundered Planet* said it all: people were doing the plundering, and the scale of degradation was global. It was also a vision of the consequences of actions in a world where "each part is dependent upon another, all are related to the movement of the whole." "Man" had inherited the Earth, but was now wreaking "havoc . . . upon his natural environment." Osborn went so far as to proclaim that "today one cannot think of man as detached from

the environment that he himself has created." Yet this was not a question of maximizing the efficient extraction of the necessary resources
for civilized life. Instead, how humans behaved would determine
their chances of survival in an age when humankind was "becoming
for the first time a *large-scale geological force.*" "One wonders," pon-
dered Osborn, "what obligations may accompany this infinite posses-
sion."[15] Those who invoke the term *Anthropocene* in the twenty-first
century to draw attention to much the same problem may justly wonder, on reading this, why the arguments of 1948 did not translate into
a politics that could answer Osborn's question before its reformula-
tion, or reinvention, a half-century later.

Why this word *environment*? Vogt and Osborn almost certainly
did not anticipate a world of environment ministries and protection
agencies, of a United Nations Environment Program. The revolutionary career of this word was entirely an accident. And yet not. For an
examination of those influences that shaped Vogt and other contem-
poraries will show how powerfully they shaped, in turn, our own
imaginary of what "the environment" is and who is placed to talk
about it, or even govern it. And this history is closely connected with
the resonances that particular word had among many people who
heard new arguments about the planet and reconfigured their thinking about it. The history of an idea, and particularly the idea of the
environment, is also the history of a style of imagination, a history of
sciences, and very much a history of politics.

Four Dimensions of "Environment"

In this book we put forward four dimensions that together shaped
the concept of the environment in the postwar era. At first sight, this
list may surprise the reader. None of them may seem to be very much
about the environment. Yet delve a little further and you'll find that
they have become inescapable parts of how we think about it, almost
as if they are hiding in plain sight. An idea that self-evident turns out
to be the result of work, of techniques, and of its history.

The first dimension is future. In his famous though controversial
1959 lecture on "the two cultures," the Cambridge chemist and nov-

elist C. P. Snow declared that scientists "had the future in their bones." A profound orientation toward the future, and the possibilities of accurately predicting it, was a major preoccupation of postwar science and politics as capitalism and communism competed to offer a plausible vision of what people could expect next. The environment was an idea that burst into life in a futurological soup, but in this particular case it was framed by a narrative about the planet in which scientists both identified a general and advancing degradation in the world around them and felt it incumbent upon themselves to provide solutions.[16]

Our second dimension—expertise—was a means to try to identify and adjudicate between these possible futures. This was a period when scientific expertise itself was shifting, especially in those areas with environmental interests, as the core skill prized in research leaders became less about fieldwork conducted in particular places, and more about the *processing* of information gathered from multiple places. The new expertise required a capacity to consolidate information into models and data sets at scales beyond anything an individual collector could achieve. Among the latter we include those who could handle or generate "big data" and the new institutions in which they worked, increasingly in tandem with government.

The emergence of this processing expertise worked in a kind of feedback loop with the idea of the environment itself. "Environment" was an "integrating" concept that worked across traditional scientific disciplines and demanded new ways of collecting and handling information. This was part of its power and attraction. This change in expertise must be understood as a *relative* shift, and one that took place at different rates in different disciplines, drawn out over decades, but one that also often altered the primary purpose of fieldwork and measurement. The focus moved away, for example, from searching for species in an ecosystem in order to study them and understand their evolution or function toward treating them as "indicators" of total environmental health, and creating new kinds of scientific controversy over issues such as measurement quality or risk thresholds.

These developments must also change what we might traditionally think of as the history of an idea. When the idea becomes associated with the work of thousands of researchers, with government programs and international diplomacy, the very substance of the idea is stretched and molded by their practices, the conferences and workshops, the diplomatic compromises. We will see this repeatedly in the history of the environment. The long and deep thinking of individuals and their˙moments of inspiration and articulation remain part of the story, but less so over time. Ideas are not just shaped by lone people or collections of women and men in smoke-filled rooms, but just as much in conference halls and laboratories where people come in serried ranks.

Processing environmental information increasingly became wedded to a set of techniques integrating numbers and using computers. This idea was not just about people and minds, but machines, especially at scales of the environment beyond easy observation (of which an object of study such as the "global climate" is an obvious example). The idea went virtual, and what is thought to be important about it is shaped by the work computers can do. Having the means to do this processing and comprehend the environment as a complex and interconnected web made authoritative expertise more "aggregative," culminating in the multiauthored mega-reports of international bodies, typically making major claims about the future. In this book we will talk about "aggregated expertise." We will see environment's extraordinary elasticity, stretching from Aldo Leopold's "Land Ethic" to the latest report of the Intergovernmental Panel on Climate Change. Yet over time it became the aggregate of techniques and institutions that shaped the idea more than individuals. This also affected who could speak up for the environment with political effect—and who had power over it.

This leads us to our third dimension: trust in numbers, a phrase borrowed from a work by Theodore Porter on the development of nineteenth-century social science, where he describes how quantification permitted the conceptualization of society as an engineering project and helped make it legible and, above all, predictable.[17] Numbers performed an equivalent function for the environment, sometimes presented in striking graphical illustrations. Were things getting better or worse? Were the graphs going up or down? In turn, over time, this lent a particular prestige to experts in "Big Science" who could generate such numbers, often on a very large scale with projects such as the International Biological Programme (IBP, 1964–74) or the United States' National Science Foundation project for Long-Term Ecological Research (LTER, first iteration: 1982–86).[18]

The capacity of numbers to indicate change was their essential contribution to this expertise. The environment as a concept was, after all, born precisely out of the idea that things changed and that the change was caused by humans. Numbers provided a trajectory to this story and permitted apparently more precise readings of the future. Politically, having numbers at hand became an essential underpinning of plausibility and authority in making a case for policies. Although never in truth detached from rhetoric and narrative, numbers nevertheless gave the appearance of objectivity and neutrality; they could be translated between disciplines and were suitable for modeling.

At the same time, the environment was an idea that linked the very local, or even the microscopic, to a planetary whole. Thus the fourth crucial dimension was scale and scalability. Part of the power of the concept of environment was that it was *already* familiar to people who worked on very different scales but who had not imagined all the dimensions it would later achieve. We will see this at the beginning of the popularization of the term in the English language through the writing of English polymath Herbert Spencer, who delivered a heady brew connecting psychology, sociology, ecology, and evolutionary theory in the second half of the nineteenth century. His thinking sought integrated ways of thought that encompassed the workings of the mind and the tiniest species but also society and nature more broadly. This made it a simple matter—we might even say an expectation—to suggest that environmental matters were intimately connected from the smallest scales to the planetary. An argument about the planet had local, microlevel implications. But the reverse might also be true. It helped that by the 1920s and 1930s the concept of environment had been adopted by groups as diverse as scientists working in cellular biology, urban public health professionals, geographers describing the habitat of "nations," and ecologists analyzing the habits and ranges of species. Now they became integrated, not just in name but also in sometimes employing common statistical techniques that could be used on very different scales (logistic curves, computer modeling). Equally, some of the new phenomena being described in the postwar period, perhaps most notably radiation from atomic bomb detonations, reached over great distances but were manifest in small, localized, indeed intimate forms: from the nuclear blast on a Pacific atoll to the baby teeth thousands of miles away that contained isotopes from its fallout.[19]

These four dimensions worked in combination, giving a distinctive form to the postwar environmental turn. How that happened was neither intentional nor perhaps predictable. The revolution that ensued was accidental. Thus was created a new and particular environmental expertise, an expertise for the future. And as the natural world became seen as an increasingly integrated and systemized entity, so did those who measured and modeled it come to be seen as, or act as, a "voice of nature" themselves.

The Future and Its Expertise

By the early 1960s, "the environment" was emerging as a potential area of government policy, although this was not institutionalized until the end of the decade or later. A leading advocate was American urban planner Lynton Caldwell, who published the seminal "Environment: A New Focus for Public Policy?" in September 1963.[20] Caldwell opened the piece with a description of a traffic jam on a Los Angeles freeway: how might one think about this? Different experts thought it represented different kinds of problems, depending on their background. One might consider it as an issue of congestion, another of air pollution, or engineering, municipal government, or finance. The problem was, he argued, that these visions were partial, considering only one aspect of a complex whole.

It may be that our failure to cope adequately with certain large and complex problems of our time is a consequence of failure to see the unifying elements in the complexity. In our characteristic concentration on intensive, specialized analysis of our public problem we may omit so many data from our normal field of vision that the integrating profile does not appear.

The purpose of this article is to ask whether "environment" as a generic concept may enable us to see more clearly an integrating profile of our society.

Caldwell went on to discuss the dilemmas of managing competing interest groups and the practice of trying to place valuations on nature. He noted a particularly problematic aspect of the emerging idea of "the environment," largely shaped as it was by discussions among scientists trying to bridge traditional gaps in approach:

In shaping our environments, we have seldom foreseen the full consequences of our action. The more remote and complex results of environmental change could not be perceived without the aid of a scientific knowledge and technology that we are still in the process of creating. . . . there appears to be no clear doctrine of public responsibility for the environment as such. It therefore follows that concern for the environment is the business of *almost* no one in our public life.

At roughly the same time, across the Atlantic in Britain, the tireless scientific advisor Solly Zuckerman was trying to prod various scientific disciplines toward increased cooperation, inspired by the example of the International Geophysical Year of 1957, which had successfully encouraged much international collaborative research, particularly on the oceans and polar regions across the icy frontiers of the Cold War. Zuckerman was from South Africa and was a primatologist by training but had established a reputation in government circles in Britain during the war as a pioneer of operational research and military strategy. In 1959 he coined the term *environmental sciences* in a memo and later played a leading role in establishing the United Kingdom's Natural Environment Research Council (NERC) in

1964, its chief means of funding environmental research. In fact, none of the participating scientific disciplines, variously preoccupied with the oceans, land, and atmosphere, had sought to put *environment* into the name of this funding body. But each had rejected alternatives suggested by others, and so *environment* was in fact a compromise: the universal second best. Environmental science was integrated by default but no less influential for that. NERC remains the United Kingdom's "leading public funder of environmental science."[21]

By this time, "environment" may have found a reception among academics, but it was also becoming a *political* concept that both demanded an institutional reconfiguration of how the natural world was studied and understood, and was coming to drive certain policy agendas. By the mid-1960s, the environment was creeping toward the policy mainstream.[22] The year 1970 saw the founding of explicitly environmental ministries in the United Kingdom and France and the Environmental Protection Agency (EPA) in the United States, all demanding environmental experts. Lynton Caldwell had been instrumental in drawing up the remit of the EPA. Many of the fields of action they dealt with were old and familiar: pollution, conservation, deforestation, and public health. Alongside such institutional developments came the rise of the popular social and political movement, environmentalism. By the 1980s new themes had moved to center stage, such as biodiversity loss (see chapter 4) and climate change.

The story of the environment we tell is not so much about *whose* idea it was or *where* it was born but *how it was made* and how it historically became the responsibility of certain branches of government, particular kinds of experts, and perhaps society as a whole. By the end of the 1960s, the environment was emerging as a standard national policy area in many places at once. In 1965 the US Environmental Pollution Committee delivered a report to President Lyndon B. Johnson on "Restoring the Quality of Our Environment" (to which we will return) that focused on pollution but also embraced public health, potential ecological effects, impacts on soil and water, and even possible climate change driven by carbon dioxide emissions. That same year, the American Association for the Advancement of Science's

Committee on Science and the Promotion of Human Welfare observed, "The entire planet can now serve as a scientific laboratory."[23]

Numbers helped: government reports were filled with numbers showing the world was in trouble—as if something so big couldn't be in trouble all at the same time unless there were numbers to prove it. Indeed, while the identification of environmental problems also opened whole new areas of field research, by 1950 a more general problem in the sciences was becoming a superfluity of data and a lack of the means to easily digest it and make it intelligible. The computer helped fill this essential niche. The analytical possibilities presented by computers offered the opportunity of *simulating* an ecosystem or a climate even when data remained far too scarce to build a rigorously empirical view of global dynamics. Computer-assisted data sets fed demand for better global models, finally building such immense amounts of information that only computers could assemble and analyze them. Carl-Gustaf Rossby was one of the key pioneers leading computerized Numerical Weather Prediction, the first example of which he supervised in Stockholm in 1955 (see chapter 5). It could be observed of him, possibly as a virtue, that he never made an actual weather observation.[24] His own expertise was largely synthesizing and theoretical, but he was at least as much a master of communication and interaction; he organized skillfully the works of mathematically talented collaborators, such as Jule Charney, who over many years built the computers, assembled the data, and ran the tests with their mathematical skills. The 1950s work of Yale Mintz at the University of California, Los Angeles (UCLA), to develop an early general circulation model (GCM) of the atmosphere and oceans was characterized in an obituary as "heroic efforts . . . during which he coordinated an army of student helpers and amateur programmers to feed a prodigious amount of data through paper tape."[25] In the data-rich age, heroism was redefined managerially: it demanded the capacity to endure an extreme level of mind-numbing tedium fiddling about in an office.

Thus the postwar period increasingly saw environmental expertise becoming detached from field or laboratory science, expressed instead in mastery of integrative and comparative techniques. Mathematical techniques and computing rather than biology led this revolution. Inevitably, perhaps, expertise also became aggregative, in that no one person or even institution could easily collect, process, and analyze the range of data required to make observations about the environment as a global whole. This had important consequences. Charismatic leaders did not disappear, but their "genius" was no longer individualistic, as it was in the time of someone like John Tyndall (1820–93), the mid-nineteenth-century Irish scientist who developed theories of atmospheric chemistry and wowed audiences with public performances at the Royal Institution in London and during a lecture tour to the United States. Increasingly, charismatic leadership demanded *representativeness* and consensus. For example, climate scientist James Hansen could claim to be speaking for his whole profession when summoned to US Senate hearings on global warming in the long hot summer of 1988, as fires raged in Yellowstone National Park. Staging and showmanship were marks of authority for both men, but their referent shifted. The new scientific politics related to the increasing presence of experts on consultative committees backed by government funds flowing into research and multiauthored seminal papers in *Nature* or *Science*, laying out how we must understand the way "the planet" is going and with it "humanity."

The virtue of aggregative expertise, the apogee of which is the Intergovernmental Panel on Climate Change, established in 1988, is the apparent anonymity of its production. The output of thousands of individuals adhering to process becomes the guarantee of its accuracy. In fact, the politically acceptable final output may be an averaged view of a range of scenarios offered by the numerous participants within it, the end results being rather distant from the models or data each of those individuals use. Personal responsibility for the final output is increasingly murky, as each participant contributes to "the scientific consensus." The *infrastructure* of prediction (institutions, conferences, computer technology, and the like) becomes crucial to the process yet is exposed to error in any one of many integrated metrics and datasets. Practitioners use code in models whose origins may be quite opaque, relying on "craft skills" in translating observations into numerical data series to fit the model. This "normal" practice of science may appear to laypeople to contradict the virtues of both modeling and aggregative expertise.

All these experts, all these billions of dollars, all these globe-spanning institutions, are a lot of weight for one little word—*environment*—to carry. Yet the curious history we tell here reveals how they were enabled by it and how that word inexorably seems to lead back to these global institutions.

The environmental outlook of 1948 was first and foremost a work of integrative imagination, of combining a set of already-existing issues and problems into new meaningful wholes. Vogt launched what might be called the "modern environmental problem catalogue." It included, but was not limited to, population growth (by far the number-one issue at the time), water scarcity, soil erosion, overconsumption, overgrazing, overfishing, pests, industrial wastes, the retarding productivity of soils, and species loss. None of these was entirely new, but they were brought together into a quite particular way of perceiving the world, a way of using scientific facts to establish what might be called a "survivalist agenda." The very word *survival* is important—not only because it appeared in the title of Vogt's book but also because it evokes a historic moment of survival after the most comprehensive war the world had seen. Survival now came to frame the human predicament itself. It was no longer a matter of the survival of the individual or the nation or a single species or a place of natural beauty—what was arguably at stake was the survival of humanity in its entangled and deep relationship with nature.

"The environment" was a crisis concept, born out of a sense of urgency in dealing with looming challenges of unusual magnitude. But it was also, paradoxically, a concept grounded in the middle of postwar reconstruction, so it was a concept of peacetime. A new era was in the making. The war had been total and global. The new world would also be global, and increasingly postcolonial. In 1948, following the UN's plan for the partition of Palestine, a broad secessionist movement had already gained momentum across the European empires, with India, itself partitioned in 1947, as a forerunner. Within little more than a decade most countries of Asia and Africa were independent states. It would also be a fossil-fueled world. Iranian oil fields had been critical during the war. In 1947, the great Saudi oil fields were discovered. Vogt's was a pessimistic message in an age both optimistic and bruised. It was a kind of schizophrenia built into the concept of environment, or survival, parallel to the contemporary message about the virtues and vices of atomic power.

With time, environment absorbed the energies of numerous intellectual and scientific strands in a way that no other concept had the capacity to do. Its predecessors (conservation and preservation) did important groundwork, and its later followers and contenders for conceptual space and influence (sustainability, ecological modernization, ecosystem services, Anthropocene) have made valuable contributions. But none of them has so far proved flexible and malleable enough to productively harbor the many tensions and contradictions that are embedded in the most recent phase of the human-nature relationship.

The environment was nobody's intention. It entered the writings of 1948 almost as a virus, percolating into minutes, agreements, plans, and pamphlets. Nowhere did it appear in a title; nobody called themselves by that name. There were no environmentalists to be found loitering or campaigning, no departments of the environment in government offices, no schools or institutes of the environment in universities or think tanks. The revolution of "the environment"—the *conceptual* revolution, as distinct from the political transformation connected with the environmental social movements and the struggles over environment—was silent, unsought, and largely unheralded. Yet without it, the world would have traveled in other directions. The environment emerged with a new unifying power. It was an idea whose time had come.

Expertise for the Future

Man and One Woman

The world was (and still is) full of earnest conferences and assemblies. Some of them dare to deal with the whole globe. Yet none before matched the ambition of "Man's Role in Changing the Face of the Earth." It was the brainchild of William L. Thomas, director of the Wenner-Gren Foundation for Anthropological Research, which provided the funding. Thomas recruited three extraordinarily influential professors to lead the event held in the middle of June 1955 at the Princeton Inn, New Jersey. Geographer Carl Sauer, entomologist Marston Bates, and polymath historian, philosopher, and planning theorist Lewis Mumford, all three American, presided over a mix of academics, policy makers, and figures from industry. The conference gathered seventy-three researchers who together offered big thinking for troubled times, albeit done among the quiet avenues of a scholarly town in stolid, wainscoted parlors. The pace was leisurely, allowing time to reflect and recline on the inn's verandas, taking in the calming views over expansive, trim lawns.

The conference was organized in three parts, covering the past, the present, and the future.[1] The centerpiece was the singular phenomenon: "man." The gathering was intended as a kind of audit of how humanity had shaped its world and its future trajectory. Only one of the seventy-three participants was a woman, the Indian plant geneticist Janaki Ammal, who had recently returned to that country from Britain at the invitation of Prime Minister Jawaharlal Nehru to lead the Botanical Survey of India. It was almost entirely men who gathered to pronounce on the influence of man.

What kind of expertise did you need to assemble to find out how "man" was affecting the whole Earth? And covering the whole of human history? If we break down the attendees assembled by Sauer, Bates, and Mumford into their disciplines we get an early glimpse of what the organizers thought constituted "environmental" expertise in the mid-1950s. Or, put another way, we see who was playing a role in actively defining what "the environment" was going to be and who claimed knowledge about it. Forty percent of the attendees came from the earth sciences, 28 percent from the biological sciences, 12 percent the social sciences and humanities, and 20 percent from applied fields such as planning. They were an elite group predominantly made up of natural scientists, with a smattering of social scientists and humanists. Doubtless the organizers, in casting a wide net in the hope of capturing knowledge about the whole planet, hoped to be as eclectic and wide-ranging as possible. Yet they had already made choices about who really counted. Tacitly, and through their interaction, they identified what kinds of facts about "the changing face of the Earth" were going to be important. Of course, this was only one meeting, and there would be many to come. Environmental expertise was not forever defined during those pleasant June days. The point to remember is that facts and knowledge considered relevant to "the environment" were powerfully shaped by meetings such as this.

The anxious and animated gathering at the Princeton Inn was not, as yet, explicitly "environmental." The word was not employed much. Activists and intellectuals still regarded themselves as "conservationists," not "environmentalists." As a kind of presiding saint, the organizers evoked George Perkins Marsh, American writer and diplomat, who had published his seminal *Man and Nature; or, Physical Geography as Modified by Human Action* in 1864. Marsh's work became a lodestone for those arguing that human behavior was damaging the environment (a term he did not use). It provided a narrative that "man" was on a trajectory toward disaster if he did not change his ways, a story historians often call "declensionist." One of Marsh's major concerns had been damage to the soil, the foundation of agriculture and civilization. His volume stood alongside other nineteenth-century writings that explicitly linked the fate of civilizations to their management of the soil.[2] Nature, which had overawed and shaped the long evolution of *Homo sapiens*, was becoming a source of evidence that something was wrong with man. Almost a century after Marsh wrote, the idea of a problem in the relationship between human beings and nature was being transformed into a story about the environment.

The explosive growth of environmentalism as a social, political, and media phenomenon in the 1960s and 1970s did not emerge from nowhere, as if scales had fallen from a generation's eyes. There had been (at least) a century of preparing the ground and developing the language through which the activists of the baby boomer generation could articulate their concerns. But in contrast to the work of Marsh, the vision of a single, polymathic individual had become inadequate to grasp what was happening to the world. The environment needed more: an aggregate of experts, a team. The history of the word is also a history of moving from the lone genius in his study to a work that demands conferences, institutions, a great collective.

Environment before *the* Environment

Environment was borrowed from French into English in 1827 by man of letters Thomas Carlyle (1795–1881). Ironically, given the fears of humans as a dangerous, growing population as expressed by writers such as William Vogt, Carlyle employed the concept of environment to understand the unique and unrepeatable character of great thinkers. The problem was not to predict or restrain "man" collectively or as a total population but rather to explain his singular genius. Carlyle used the term in the classic way "environment" was employed in that century before *the* environment. It was used to describe those extrinsic conditions that shape the real object of the study or story—frequently an individual person, but later also species, a nation, or a race. This was still going strong in the 1930s when Isaiah Berlin published *Karl Marx: His Life and Environment,* a book that certainly would not carry such a subtitle today.[3]

How then did we get from a word used to explain the lives of a single individual and used by a grand essayist and biographer expressing his singular thoughts to a world of conferences, "aggregated expertise," and *environment* as the name for, well, everything on the planet being connected? This was a remarkable transformation. Yet it took time and had two significant and rather different phases. In truth, Carlyle's usage had little initial impact. The *Times* of London, Britain's leading daily newspaper, employed the word only one single time during the whole of the 1840s. In the 1850s, it fared no better. Then things began to change. The real popularizer of the term however, both inside and outside the world of academic study, was not Carlyle but Herbert Spencer.

Spencer was born in 1820 into a similar, if less illustrious, background to that better-remembered Victorian and evolutionary thinker, Charles Darwin—the radical, well-to-do classes of the manufacturing Midlands. Spencer was an extraordinarily dynamic, ambitious, powerful, and light-footed thinker, perhaps too light-footed to leave a clear-enough trail of evidence in his arguments for many twentieth-century minds, and his star fell rapidly after his death. The star has fallen further, as more recent generations noted the racist and eugenic implications of much of his thinking. But he impressed hugely at the time, being a talismanic figure in evolutionary thinking, sociology, psychology, and political economy. For Spencer, "Progress . . . is not an accident, but a necessity. Instead of civilization being artificial, it is a part of nature, all of a piece with the development of the embryo or the unfolding of a flower."[4]

Spencer moved to London in the 1840s and entered the debates of political economy, working for the recently founded magazine the *Economist*. In the middle of the nineteenth century this intellectual world was presided over by John Stuart Mill, whose economics and liberalism were deeply imbued with an interest in psychology. As was quite typical in the economic milieu of the time but largely forgotten in the twentieth century, thinkers in this tradition saw political economy as being grounded in patterns of social behavior and psychological processes. A theorist of society thus had to be a theorist of the mind, and when the mind was shaped by physiological processes, so must biology play a prominent role—according to Spencer.

In his first book, *Social Statics*, in 1851, he presented the world as a constantly shifting, dynamic place with an essential unity of natural and social processes. "Every age, every nation, every climate, exhibits a modified form of humanity; and in all times, and amongst all peoples, a greater or less amount of change is going on . . . all evil results from the non-adaptation of constitution to conditions. This is true of everything that lives. . . . nor is the expression confined in its application to physical evil; it comprehends moral evil also."[5] Spencer had a profoundly evolutionary view of the world. In 1851 he still used an old word that unified the "external conditions of life": *climate*. But while debates about climate change had been quite prominent at moments in the eighteenth and early nineteenth century, it did not sit easily with the restless world that Spencer imagined. He was profoundly influenced by the embryology of the German-Estonian Karl Ernst von Baer; like Darwin, by the theories of Thomas Malthus (see chapter 3); by the ideas of liberal political economy; and by the evolutionary theory of the French naturalist Jean-Baptiste Lamarck (1744–1829). Taking his cue from Lamarck's theories, in which characteristics developed during the lifetime of an individual could be transmitted to his or her descendants, Spencer saw both natural and social processes as improving "fitness" and tending toward perfectibility. He thus remained skeptical of Darwinian natural selection (as did many others) and the notion that only random variation could transform species and determine "fitness" in the struggle for existence.[6] For Spencer, an understanding of society was a kind of "transcendental physiology," with society itself the "social organism," as he wrote in 1857. Of course, the idea that human society and natural processes could be studied using the same techniques and within the same explanatory framework was hardly new but had rarely been supplied with a convincing method of analysis.[7]

Spencer's work did not, in the end, provide a unified or universally adopted method. Yet this ambition and the very wide reception of his evolutionary theory facilitated the proliferation of the term *environment*. This led to a multiplicity of meanings being applied to the word, paving the way for a wide reception of its use in the second half of the twentieth century. It was not some bolt out of the blue; it was a term that everybody thought they had some handle on already.

Spencer, insofar as we have discovered, first uses the word *environment* in print on page 194 of his monumental but ill-received *Principles of Psychology* (1855). The environment is simply the source of stimuli that produce sensory effects in the mind: "those properties of things which we know as tastes, scents, colours, temperatures, sounds, are effects produced in us by forces in the environment." This is revealing. The environment is not a thing; it is what a body or mind senses to be "external circumstances"[8] that act upon it. This notion could thus be rescaled to describe the relationships driving evolution in species or societies that Spencer had described in *Social Statics*. He became most famous for his sociology, and it is in the application of *environment* here that it won its broader application to notions he had already formed about the "universal laws" that drove nature and society. In *The Principles of Sociology* he wrote, "on . . . conditions, inorganic and organic, characterising the environment, primarily depends the possibility of social evolution." But then one must consider factors "which social evolution itself brings into play," such as "the progressive modifications of the environment . . . which the actions of societies effect," meaning climate change, deforestation, intentional or unintentional movement of species, drainage, and so on. Yet while Spencer's usages here may seem strikingly modern, the environment is still no more than a name for a list of attributes to which an evolving society is subject; it is not an integrated system, any more than what Spencer called "the super-organic environment" was, meaning "those adjacent societies with which it comes on the struggle for existence" that in turn determined the precise form of a society's government.[9]

Spencer's legacy is primarily remembered in sociology and what was later labeled "social Darwinist" politics, and thus generally he is not recalled sympathetically. His harsh conclusion from his belief in perfectibility through adaptation, bringing improved fitness in a dynamic world, was to insist on laissez-faire economics and absolutely minimal welfare provision. Anything else breached the free exercise of the faculties required to generate fitness. Drawing on his inspiration from embryology, he saw processes of "differentiation" and the emergence of heterogeneity from homogeneity as the primary characteristic of a progressive evolution. He then applied this, clumsily, to racial theories of the day, seeing social development and speciation as essentially the same thing: the bushmen were analogous to protozoa, aboriginals to polyps, tribal nations to hydra, early commercial societies to annulosa or crustacea, and so on.[10]

The legacy of Spencer in this history is both the use of the word *environment* and habits of thought among later authors, mimicking him as the master, moving easily—and sometimes rather superficially—between biology, psychology, and sociology. The 1933 English classic *Culture and Environment* by literary critics F. R. Leavis and Denys Thompson understood *environment* in a tradition that drew on the biographical view of Carlyle, as the conditions that shaped "training taste and sensibility" (as they saw it). But they also drew on a sense of environment as the envelope for an entire way of life. English writers already frequently employed the term *environment* in reference to Thomas Hardy's writings in the 1910s and 1920s, again addressing writing that melded character and landscape.[11] Hardy himself rarely used the term, and he usually used it only in a very general sense of something that shaped "character."[12] Leavis and Thompson contrasted their own efforts as scholars with what they saw as the deleterious cacophony of modernity, "multitudinous counter-influences—films, newspapers, advertising—indeed the whole world outside the classroom." They wrote of the "wanton and indifferent ugliness . . . of the towns, suburbs and houses of modern England" as an affront to the "natural environment." They thought "ugliness" was most distinctive of the modern environment, contrasting it with a vanished "organic community" whose demise had taken with it "human naturalness or normality."[13] As a portrayal of the changes attendant on industrialization, this was doubtless little more than lazy stereotyping, but it expressed a sensibility frequently found in the twentieth century: that nature can hardly be uttered without evoking some nostalgia, a narrative of loss, while *environment* is more a statement of fact. This too paved the way for the reception of a new idea of the environment after 1948.

This "geographical" and "character-forming" tradition could be found in the academic work of the American Ellen Churchill Semple and Australian T. Griffith Taylor (author of *Environment and Nation* of 1936), both advocates of what we now call environmental determinism from the perspective of the colonies. The experience of colonial expansion and ethnographic encounter had long raised questions about whether different geographies produced different kinds of people.[14] They drew too on ethnographical and geographical traditions, such as the writing of nineteenth-century German thinkers Wilhelm Riehl and Friedrich Ratzel (the latter trained as a zoologist and had an interest in Charles Darwin and Ernst Haeckel). Retrospective twentieth-century English translations of these German writers sometimes use the word *environment*, but they did not use any such consistent term themselves, and the modern German *Umwelt* had not yet acquired its modern meaning as a translation of *environment*. These German geographers explained an organic, rural society through its attachment to the land and saw urban cosmopolitanism as severing these links.

These works were widely read but often rather idiosyncratic. They indicated a broad mood favorable to evolutionary theory rather than the emergence of a new *discipline* wedded to the term, with its conferences, departments, and armies of cross-referencing researchers. Amid this ferment, however, "environmental" did become an increasingly important term within the nascent science of ecology. Prominent early ecologists in America such as Stephen Forbes (1844–1930) and Frederic Clements (1874–1945) drew directly on the ideas of Herbert Spencer in formulating their ideas of a holistic community of species, although as we shall see in chapter 4, there was certainly not only one way to do ecology.[15] However, for a long time (as now), *environment* was used in the sense of "surroundings." Most ecologists focused narrowly on the physiology of particular plants or small associations of them. The ecological insight—that these surroundings somehow might form part of an interconnected whole or system—only emerged slowly. The narrative was about what the environment did to the hero (human or nonhuman) of the story, and it was only interesting insofar as it produced effects within the hero. There was no story to tell *about* the environment. This is implicit in the early history of ecology. To use one example, William H. Howell wrote in 1906 of how "the varied and important reactions between the organism and the environment should be included under ecology." Like Haeckel, the German polymath who coined the term *Ökologie* (ecology) in 1866, he was writing about the relationship between physiology and ecology, and environment was interesting because of the light it shed on physiological questions.[16]

Spencer sought to combine thought about society and nature, and it was an approach many others tried to emulate. One was a young American botanist, Lester Frank Ward, who picked up on the restless dynamics of Spencer's vision and in 1876 used it to attack notions that the distribution of plants and animals was stable and settled. For Ward, evolution was a struggle for place where each species had "a potential energy far beyond and wholly out of consonance with the contracted conditions imposed upon it by its environment." Here environment was not a restraint but the challenge against which progress stepped up and measured itself.[17] Ward would later follow a Spencerian trail, though not Spencer's conclusions, by becoming a pioneering figure in American sociology.

Another profoundly influential writer was the Cambridge professor Alfred Marshall (1842–1924), one of the most important figures in shaping the modern discipline of economics. He was determined to reshape his subject as a proper science and saw evolutionary theory as the framework by which economic development could be understood over time. Marshall used environment frequently and almost always as a metaphor drawn from Spencerian evolutionary thinking, but also as an attempt to translate the German term *Konjunctur*.[18] The American theorist Thorstein Veblen also used environment frequently in his famous *Theory of the Leisure Class,* a book on the boundary between sociology and economics. Veblen sometimes directly invoked Marshall as a context in what he called the "struggle for existence" of social forms and businesses. Veblen was engaged in continued late-nineteenth-century debates between Darwinists and Lamarckians over evolutionary theory. Veblen rejected, however, the possibility of physical human evolution on a short historical time-scale, which Spencer saw as being possible as social forms evolved. Instead, Veblen concentrated his idea of evolutionary change on institutions that as a "social environment" shaped habit, possibly influenced by the British biologist C. L. Morgan.[19] In all uses at this time, the term *environment* bore the sense that interior life was captive to exterior forces. The environment, then, was that part of the exterior world that became interiorized in that it produced lasting effects on physical or mental life: "Thus it becomes morally certain, that, at last, great, general, permanent, and all-important facts in the environment, will produce in organisms impressions so deep and lasting that they will tend to become intuitive and instantaneous," as a presentation to a Unitarian conference put it in 1866.[20]

A return to the *Times* of London can help us gauge some of the word's newfound popularity. Used once a decade in the 1840s and 1850s, *environment* hardly fared any better in the 1860s, with six appearances. By the 1880s it was up to 115, and by the first decade of the twentieth century, 733. Now one might expect some broader familiarity among the readership. This reached 1,455 mentions in the 1920s, a level that remained roughly static until the 1960s when 4,746 was reached (of course, over this whole period the newspaper was also getting bigger). All these uses, we might say, were "individual," both in the sense that individual authors used the word in many different ways, and also that the environment being described was that of a particular thing. Yet at the same time, many thinkers now recognized the utility of the idea. It was a notion you could have a conversation about with people in far-flung and different areas of work. It was also a notion that had been particularly valued among those who sought to bring together thinking about the social and natural worlds. This was important groundwork for what would come later: first a sharing of knowledge between different experts and then attempts to create an expertise that was fully integrated.

The Politics of the Unpolitical

Vogt was not alone. Postwar reconstruction was a febrile period for the building of institutions and philosophies and thinking on a global scale. The year 1948 saw the founding of what was arguably the first environmental organization, the International Union for the Protection of Nature (IUPN). It was the year that articulated the idea of a distinct and absolute limit on essential resources, which would later be captured in expressions such as "peak oil."[21] At the other end of the spectrum, 1948 was the year of implementation of Joseph Stalin's "Great Plan for the Transformation of Nature" in the Soviet Union, responding to the 1946–47 disaster of drought and subsequent famine that resulted in half a million deaths. A series of dam and irrigation projects, designed to protect the future of agriculture on the Russian steppes and plains, ultimately wreaked new havoc, including the desiccation of the Aral Sea.[22]

The Soviet program, modeled on the more modest American response to the Depression and Dust Bowl of the 1930s, signaled that in the East, as well as in the West, mainstream thinking was directed toward constructing a nature to be set to work to benefit humans rather than protecting and managing a nature under threat. And nature was soon to be transformed more rapidly than ever by rapid (sub)urbanization, by the rise of motor vehicles and diesel tankers, by unprecedented economic growth and efforts toward economic development in a swiftly decolonizing "South" by both the capitalist and communist blocs of the "North." This global South would soon be termed the *Third World*. The environment came to be a counter-concept to many of the consequences of this kind of mainstream modernization and, later, a rallying cry for those who locally resisted developments, often imposed by outsiders, with major social and environmental impacts. Yet in its inception, the intrusion of the environment into politics did not come from a marginalized position.

Instead, it spoke directly to the heart of the scientific and political establishment and soon became a flagship issue for these elites.

Rebuilding the war-torn world was on one level a straightforward undertaking. Homes, roads, railways, factories, electric and sewage facilities, and other infrastructures had to be built, repaired, and replaced. The economy of war had to be transformed into an economy of peace. And like the economy of war, which had created an unprecedented level of scientific involvement in government, the postwar order needed droves of experts, working together. Staff was needed to run companies, institutions, government offices, local administrations.

Amid these largely national efforts, the newly constituted United Nations undertook work on "conservation and utilization of resources." The official initiative came from President Harry S. Truman through the US representative on the United Nations Economic and Social Council in September 1946. In 1948, preparations had matured enough for the United Nations to issue a seven-page memorandum with a description of the background, rationale, and general layout of the United Nations Scientific Conference on the Conservation and Utilization of Resources (UNSCCUR). President Truman pointed out that natural resources had demonstrated their crucial importance during the war and could be considered a major cause of conflict: "The real or exaggerated fear of resource shortages and declining standards of living has in the past involved nations in warfare. . . . Conservation can become a major basis of peace."[23]

There was widespread agreement that natural resources had been important for the war effort, and the war itself had destroyed them in many places. To tackle the availability of resources across the world, Truman envisaged "an exchange of thought and experience" among experts who would "not necessarily represent the views of the government of their nations, but would be selected to cover topics within their competence based on their individual experience and studies."[24] The program of the planned conference also touched on nature conservation. In fact, these parallel strands of thought were reflected in two conferences held next door to each other, both under the auspices of the United Nations. Sessions were scheduled to enable delegates to UNSCCUR to attend a concurrent conference on nature conservation, named the International Technical Conference on the Protection of Nature (ITCPN), organized by the United Nations Educational, Scientific, and Cultural Organization (UNESCO). Equally, the ITCPN was scheduled to permit attendance at UNSCCUR. Already a difference was manifest between conserving nature as a resource and conserving nature for itself. This reflected an older division, understood in the United States as the difference between a *conservation* that ensured adequate resources for the economy and *preservation* that looked after treasured landscapes and monuments. At the same time, links were apparent, and to some degree the conferences engaged the same people.

The parallel conference was the product of a decision by UNESCO to mandate a new organization, the International Union for the Protection of Nature (IUPN) at an October 1948 meeting in Fontainebleau, France.[25] The opening mission paragraph of this new organization stated, "the term 'Protection of Nature' may be defined as the preservation of the entire world biotic community, or man's natural environment, which includes the Earth's renewable natural resources of which it is composed, and the foundation of human civilization."[26]

"Thinking globally" was by no means novel. Already in the eighteenth century the French natural historian Georges-Louis Leclerc, usually known by his title, the Comte de Buffon (1707–88), had elaborated in his book *Époques de la nature* (1778) what he imagined to be the seventh and last of the "epochs of nature." In this epoch, humanity had ascended to control nature and ruled it according to its human dictates and nature's own laws. Buffon, an aristocrat and estate owner, who had for many years been the director of the Jardin du Roi (Royal Botanical Gardens) in Paris, was used to adopting a managerial perspective on nature. He was essentially an optimist about human stewardship, despite expressing profound concern about the human tendency to destroy more than nature could repair.[27] Several intrepid and broad-ranging thinkers proposed similar ideas in the following century, although rarely on a global scale. Alexander von Humboldt (1769–1859), the Prussian polymath and traveler of the Americas, noted and sought to measure "terraforming" caused by human action: processes such as deforestation, desiccation, and local climate change.[28] But apart from Buffon's early, speculative scheme for human control, works on the power of humans to transform the Earth remained regional or, at most, and more rarely, continental. There were more eccentric versions, such as *La fin du monde par la science* (The End of the World through Science) of 1855 by the French lawyer Eugène Huzar, hypothesizing a range of global upheavals resulting from the Industrial Revolution.

But these writers did not belong to research fields or disciplines; the prestige of these aspects of their work is more retrospective, and they are now viewed as forgotten prophets of what some people call "the Anthropocene." Their concepts did not cascade through society, to be voiced by heads of state, social movement activists, urban planners, and generations of schoolchildren; neither did they attempt to analyze their surroundings as a unified, interconnected system. Very few, if any, scientists accepted the idea that humanity could change the entire Earth in any significant way, let alone change its climate other than locally. Most acceded to the argument of Nobel laureate physicist Robert Millikan in 1930 that "man is powerless to do it [the Earth] . . . any titanic physical damage."[29]

More recent inspiration came from the Russian scientist Vladimir Vernadsky's 1926 book *Biosfera* (at first accessible outside Russia in the French translation of 1929). The biosphere, defined as the critical life-sustaining zone enveloping the planet, was a new way of conceiving the world and humanity's place in it, and it meant comprehensive reconfiguring of the ways people, including scientists, thought. It was in 1948 that Vernadsky's ideas on the biosphere entered wider English-language circulation through the work of the highly respected Yale ecologist Evelyn Hutchinson. Modern "man" (as they put it) was eroding his own "survival" by wasting "parts of the *biosphere* which provide the things that *Homo sapiens* as a mammal and as an educable social organism needs or thinks he needs. The process is continuously increasing in intensity as population expands."[30]

By 1948 it was considered natural that issues of general global concern, even if not operating on a global scale, were dealt with by "modern science" and by "experts." Indeed, in addition to UNESCO, a range of additional expert bodies were involved in the conference preparations, such as the Food and Agriculture Organization (FAO), the World Health Organization (WHO), and the UN Economic and Social Council (ECOSOC), the new home base of economics in this international apparatus. Education, culture, the social sciences, and the humanities were all explicitly to be involved. However, it was the natural and technical sciences that held a privileged place in the November 4, 1948, document that finally mandated the meeting. The word *experts* primarily referred to practitioners of these sciences.

The two concurrent conferences, UNSCCUR and the ITCPN, were held at the temporary UN headquarters in Lake Success, New York, during the last two weeks of August 1949. Although the name might conjure a secluded resort, UNSCCUR in fact took place in a rapidly converted factory in a suburb on Long Island adjacent to the New York borough of Queens, as the United Nations scrambled for space in its early days. More than five hundred delegates attended the UNSCCUR from some fifty countries. Economic development was a major concern, voiced by figures such as the economists Raúl Prébisch (Argentina), Jan Tinbergen (Netherlands), Gunnar Myrdal (Sweden), and Barbara Ward (Britain). In contrast, the IUPN meeting was smaller, representing more continents but with fewer delegates from each, and focused squarely on the protection of nature.[31] Women made up to a tenth of the attendees at ITCPN while, tellingly, at the UNSCCUR women were almost nonexistent, further underscoring the latter's links with security, economics, and high politics. ITCPN was thought to deal more with issues closer to local livelihoods, traditional knowledge of nature, and the household.

While Vogt had already proclaimed that population growth and economic expansion had made the world "sick," UNSCCUR concentrated on how to use the Earth as much and as efficiently as possible. Or, as Carter Goodrich of Columbia University, chairman of the preparatory committee, told the *Sydney Morning Herald* ahead of the meeting, " 'Perhaps,' as someone has said, 'a Russian can show an American a better way to catch fish. Perhaps an Englishman can show a Yugoslav a better way to grow cabbages.'" Papers would favor new methods or technologies to squeeze more wealth out of nature, including stimulating or preventing precipitation from clouds. Australians would especially appreciate it if clouds could be manipulated to "'hop' mountain barriers" and thus travel to arid areas and release their moisture there.

Nevertheless, the alarmist narrative of Vogt and Fairfield Osborn found some reflection in the proceedings. Economist Colin Clark of Brisbane, Australia, offered a keynote paper on the general problem of population pressure and consumption on the world's resources. After special scientific sessions in the mornings, "afternoons will be devoted to plenary sessions, at which specialists representing fields as diverse as ornithology and electronics will gather. They will hear each other's views on the great overall problem—the problem of man's fight against depletion and poverty and his struggle to find new ways of improving the use and conservation of the wealth of the earth." Success at Lake Success was to be secured by it being run by scientists, with minimal interference from diplomats or politicians. Indeed, the organizers proclaimed, "Politics will play a very small part, if any."[32]

The contrast with the IUPN meeting was striking. There, protection of species was at the forefront. Delegates could listen to interventions about ecology—comprising by far the largest portion of the papers, as ecologists also made up the core of the members of the IUPN—nature reserves, new lists of endangered species,[33] toxic substances such as DDT, the idea of a sensitive natural equilibrium, the prospects of "human ecology" (much endorsed by the UN's Economic and Social Council, where humans were more highly valued than in pure ecology), the need to build public understanding, and, even, considerations of a world convention on conservation. Nature, IUPN-style, was a much more sacred and subtle thing than it was at the resource-oriented UNSCCUR.

Indeed, another biologist with a strong interest in ecology, Julian Huxley, head of UNESCO and the chief architect of the IUPN, had charged the meeting with the special mission to consider the education of the general public in the importance of nature conservation. This required considering the "means of educat[ing] the public to a better understanding of man's relationship to his environment," echoing the mission statement of the newly formed organization.[34] During the ITCPN, several speakers used the word *environment*. The concept was now emerging out of specialist and separate literatures and becoming coupled with unease at what was going on in the name of expansive, modern, industrial societies. The conference delegates did not, perhaps, employ the purple prose of Vogt, yet they did note, in the words of the IUPN secretary general Jean-Paul Harroy, "the abuses of modern economy."[35]

Thus from the very outset, a fundamental tension was built into the discourse of the environment: it was both a source of human well-being (natural resources), an "economistic" reading,[36] and also the *object*, or recipient, of the downsides of human action. Where the "waste" and "devastation" mentioned in President Truman's initiative were handled by UNSCCUR as problems to be addressed with new technologies and further modernization, development was approached rather more skeptically by the ITCPN, which cautioned against what more civilization could bring. The old distinction between "conservationists" preoccupied with resource management and "protectionists" interested in preventing human impacts on nature, familiar in American usage, re-emerged in a new guise.[37]

Yet there was also something distinctly different that was precisely offered by the new concept. The career of *environment* would also be marked by its flexibility, its openness to expansion. Its capacity to travel across boundaries and to find a place amid many different understandings and agendas, to bring people together, already gave it peculiar power in a world that had been so markedly influenced by Spencer and evolutionary thought. This power would only grow as the second half of the century unfolded. At the same time, this shape-changing power would exercise an almost magnetic attraction, pulling people and ideas into new relationships with one another. And as this occurred, understanding of the environment increasingly became a collective endeavor.

Future Environments

"In effect, the scientists proclaimed a revolution, stupendous in its implications, in the relation of man to his environment," declared a report on UNSCCUR printed in the leftist but solidly anticommunist magazine *Commentary*, founded by the American-Jewish Committee in 1945. Science could make "hunger obsolete" and the resources of land and sea adequate in the modern world, "to feed the present population of the world and any probable increases." Survival was more likely if supported "by a worldwide organization of plenty than by any attempt to retreat into autarchic, self-sufficient compartments."[38]

The "unpolitical" nature of the UNSCCUR was of course a convenient myth and one not shared by many UN member states. Most Eastern bloc states did not attend, including the Soviet Union. The Berlin airlift had ended less than a month before the conferences convened, and whatever the nonpolitical ambitions of UNSCCUR, the Cold War had politics. Fighting hunger was also another way of winning the Cold War, and the deployment of science and technology lay at the heart of that endeavor. The environment would soon become strategic. The US military launched a range of research programs on oceans, atmosphere, and geophysical conditions, stretching vertically from the fathoms of the sea to the uppermost layers of air and horizontally to extreme environments that required special knowledge for survival. By the end of the 1950s, they would be using the new meaning of the term, and the research conducted was important for building environmental knowledge.[39] This concern among the military indicates that the concept was not only related to "survival" of the planet but also to other forms of "survival" that could be rated even higher, and the concept could not be monopolized by those who wanted to "save the planet."

The environment had a politics from the beginning. Yet this wasn't only a battleground for competing ideologies; it was also clear in the relative prestige of expertise present. The smaller, ITCPN conference was clearly secondary, starting its sessions late in the afternoon to attract delegates from its larger neighbor, if they had any energy left for more discussion. It was clear which was the main game in town. At the same time, there was a sense that the division of labor among experts and scientific disciplines was not quite up to the job of what was happening to the world.

Here lay the impulse behind a multidisciplinary conference held in the wainscoted chambers of the Princeton Inn six years later, on "Man's Role in Changing the Face of the Earth." The conservationists and resource managers of the mid-twentieth century shaped the emerging environmental revolution through imagining and discussing a global problem of depletion. Put crudely, they still mostly represented the old school, where a capricious or ungenerous nature had been the enemy of "man," a danger to be conquered. But now, according to Vogt and Osborn, *man was the danger*. The jeremiads were coming from people who were from much the same milieu and meeting in much the same places as the more traditional resource-focused conservationists. Something was brewing at the heart of the scientific and political establishment.

In the words of Edmund Sinnott, president of the American Association for the Advancement of Science, speaking to that eminent assembly in 1948 (as cited in the *Christian Science Monitor*, a safe sign that his words had taken wings): "man's command over nature has grown more rapidly than his mastery of himself. Man, not nature, is the great problem today."[40] Yet the remedy was seen to lie in the very people who had developed that command over nature. They possessed the instincts and expertise to put things right.

This was the view one could find in Harrison Brown's *The Challenge of Man's Future* (published slightly later, in 1954). Brown was a scientist at the Institute for Nuclear Studies of the University of Chicago who moved to the California Institute of Technology (Caltech) as his book came out. Brown had written large parts of the text in Jamaica and drew on experiences of his time in Europe. His point of departure was largely optimistic. A fifth of the world's population lived in a hitherto unknown affluence thanks to the "machine civilization." But this civilization was, due to poor management of resources and existential threats from its own weaponry, "in a very precarious position" and was likely to "soon disappear, never again to come into existence." If this happened and machine civilization disintegrated, humanity would revert to a life which, Brown claimed, would not be unlike conditions that reigned in Europe in the seventeenth century or "in China today."[41]

Brown articulated clearly (and with a range of notable endorsements from luminaries such as William O. Douglas of the US Supreme Court and Albert Einstein) the mix of civilizational angst, technological possibility, and fear for the future of the planet that fizzed together in the years after 1948. Crucial to assessing the problem was to look to the future. "In principle, the vast knowledge we have accumulated during the last 150 years makes it possible for us to look into the future with considerably more accuracy than could Malthus." Like his predecessors, Brown claimed to see mankind in true perspective and "in relation to his environment."[42] Brown was a nuclear scientist who turned his calm but razor-sharp mind to issues previously dealt with by less exact humans.

Just like Osborn in *Our Plundered Planet* and so many of the authors in that genre, Brown strode through basic population ecology, the emergence of mankind, the psychology and biology of the species, and the long history that brought us here (in a dozen pages or so). Another characteristic of this discourse is pessimism about humanity. "Man," as humankind was called in the language of that time, is a product of evolution with very little ability to deal with complex modern social problems, let alone global ones. People are primitive and cannot be trusted to make good decisions on population, resources, food, energy, raw materials, and the consumption of goods. Because humans are the way they are—egotistic, not very farsighted, clinging to narrow group or national interests—Brown was pessimistic as to whether they could identify or implement the necessary solutions.

Where was an analysis of society here, or indeed politics? Instead, Brown sought to persuade the powers that be to adopt the perspective of this coolly rational seer—an approach entirely in keeping with that of Vogt and Osborn, in their work heading lobbying organiza-tions as well as writing impassioned jeremiads. By the end of Brown's book, he predicted a humanity that would have 6.7 billion people in 2050 and would probably not be able to survive.[43]

The prestige of science and the role of scientists in defining the environment helps explain why individuals who emerged as spokes-people for the environment, even as a kind of "voice of nature," would frequently be scientists. The society of science, with all its laborato-ries, conferences, and rituals of academic honor, provided the niche these individuals inhabited. They were not lone geniuses but repre-sentatives of a type, and they relied on the prestige of their contribu-tions to science to give them status in interacting with a wider world, often talking about things quite distant from their original training. Here we can list Vogt himself, trained as an ecologist and ornithol-ogist, who became an advocate and administrator campaigning for family planning. Other figures include Osborn (a businessman with an interest in zoology), Paul Ehrlich (entomologist and evolutionary theorist), Georg Borgström (plant physiologist), Barry Commoner (zoologist and eventually presidential candidate), Buckminster Fuller (architect and systems theorist), and E. F. (Fritz) Schumacher (stat-istician and economist). And of course there is Rachel Carson her-self, a laboratory researcher and later marine biologist.

Ten years after the Man's Role in Changing the Face of the Earth conference, in 1965, America's Conservation Foundation (with Os-born as president) convened a meeting in Warrenton, Virginia, de-liberately imitating the earlier conference.[44] Many of the same peo-ple were there. There was still only one woman—a different one, Anne Louise Strong, an American professor of planning. British ecologist Frank Fraser Darling observed from the chair that there were "a lot of ideas about the future." The ideas came from participants such as Vogt, Kenneth Boulding (economist), Raymond Dasmann (conser-vationist biologist and later director of the IUCN and professor of environmental studies), Lewis Mumford (historian and sociologist), Max Nicholson (a leading British ornithologist and conservation sci-entist), M. King Hubbert (petroleum geologist), and Lynton Caldwell (urban planner and political scientist). We meet all these people again in the following pages. Contributions abounded with comments on "public policy," "environmental impacts," and "weather modification." In contrast to the event in Princeton a decade before, the environment was everywhere in discourse, bound with a tight leash to human needs and government action. There was wide assent for the vision of Canadian ecologist Pierre Dansereau: "A valid imaginary reconstruction of our world is now our greatest task. It may even be the condition of our survival."[45] If the aggregated expertise at Princeton in 1955 had almost accidentally emerged as a "rich resource of coordinated knowledge and quickened thinking," the meeting a decade later was a purposeful "gathering of skills [to] ameliorate, prune, amplify and possibly validate the thoughts of each," as Darling put it.[46] The common thread in all the papers was "management." The delegates were part of what was becoming an established form of expertise, an aggregation brought together to manage—what, exactly? The answer lies in the title of the event: Future Environments of North America.

Resources for Freedom

Groping in the Dark

One hundred and twenty lines of computer code. That was the almost laughably small amount that went into the World1 model of global dynamics developed by Jay Forrester, working at the Massachusetts Institute of Technology in 1970. Yet those few lines of code provided the core of one of the most politically explosive academic interventions of the era: the *Limits to Growth* report published two years later to controversy and acclaim.[1] The model by that time had been elaborated into the World3 program by a small, largely American team working under Forrester's direction.[2] It predicted that economic growth was going to be limited by cost. Scarcity of resources would drive up cost, as would the expense of cleaning up excess pollution. *Limits to Growth* gave a significant and enduring boost to the idea that the economic "business as usual" of the postwar years, or even the entire Industrial Revolution, could not go on. That debate still rages between entrenched pessimistic and optimistic views. More significant was a methodological innovation, one that has won far wider acceptance—the computer that could build a simulacrum of the world and generate scenarios and predictions that would provide a lodestar for political debate—or even, in some views, determine what we should do.

Undoubtedly some of the glamour of technological wizardry explains the excitement that the report generated. The report also came out of an authoritative place. The departments, laboratories, and institutes of the Massachusetts Institute of Technology spread for a mile along the northern bank of the Charles River facing the brownstone terraces and golden-domed statehouse of the old heart of Boston. From the 1940s this institution at the heart of the old Republic was the powerhouse of the new in the United States. Focused on research and development and technological innovation, with close connections to the military, it was altogether more entrepreneurial than its more venerable neighbor up the road, Harvard. Jay Forrester had taken a leading role in developing wartime missile technology and from that moved into cybernetics, industrial organization, and an understanding of feedback systems that emerged out of wartime technologies. Mixing with politicians, computer scientists, engineers, and management theorists, he exemplified the emergence of a new kind of expertise that promised to provide more integrated tools of analysis and joined-up understandings for policy makers. In turn, the *Limits to Growth* project was inspired and commissioned by the Club of Rome, an international think tank composed of industrialists and intellectuals with close links to various governments. The title of their first prospectus in 1968, *The Predicament of Mankind*, set out the anxieties around which the group was formed, including fears about overpopulation and the environment. *Limits to Growth* was delivered with fanfare and high expectation at a launch in Washington, DC, accompanied by leading politicians, and was reviewed in the most respected American newspapers.[3] Not only did it embody the latest in the application of technology to global problems; it also helped define those problems in its own right.

The World3 model exemplified an important new variety of expertise that has been particularly important in modern environmental debates: the processor and synthesizer of data (see chapter 1). Especially in the environmental sciences, the prestige of the expert had rested at least in part on fieldwork, on getting out in the elements as a measurer, recorder, and experimenter. This was the essence in earlier times of what we will call "contributory expertise," the generation of new knowledge through a specialization in a specific scientific discipline. Of course, these people then communicated that knowledge to others, and the better they were at it (what we will call "interactional expertise") the more influence their ideas won. Such interactional skills might come from oratory or institutional author-ity (holding an important university post, for example), but could also come from adopting shared techniques familiar to people work-ing outside their own academic niche. This could win them kudos among those who applied the same techniques to other problems. But none of the young MIT team members building the World model had any direct role in gathering the information they used. And how could they, after all? How could you have experience of the whole world? Of course, as we've seen, they weren't the first people to write about the global environment. Authors such as William Vogt and Fairfield Osborn had attempted to grasp it by accumulating similar, more local cases and datasets to demonstrate that they were mani-festations of a particular process shared by the whole. The MIT team went straight to the process and then sought the data that could be put into the model to predict the outcome of that process. This was a very different *style* of working, and over the following decades modeling would prompt a hunt for new kinds of data that could be fitted to computer models.

The World3 model was extraordinarily simple by later standards, treating the whole population of the world as a homogeneous group and ignoring possibilities of political shocks or major technological shifts. This approach was selected simply because such assumptions made the modeling itself feasible.[4] The people who created the model had vast experience in developing computers, electrical engineering, and feedback models for applications in engineering, business, and social science. These critics of unrestrained growth also emerged from the heart of the postwar military-industrial complex, as did much of the new environmental thinking. This position granted them almost unparalleled interactional expertise and authority. Yet it was a new way of generating knowledge itself: a new "contributory" expertise, standing at the heart of a web of data collection but divorced from direct contact with the world.

Much of the data available on world trends in resource use or pollution was completely inadequate even for the purposes of the relatively simple model and required extensive interpolation between snippets of information. The team knew that they were pioneers launching into uncharted territory. Even a decade later, the pioneers working on computer models of global dynamics still considered themselves a somewhat embattled minority, "groping in the dark."[5] The MIT modelers understood the limits of their work. They felt that the data they could assemble was far too weak to sustain detailed projections far into the future. Nevertheless, they were confident in the *dynamics* of the relationships they could establish. Their report provided striking graphical data of the decline and collapse of the world economy under different assumptions, but they were careful not to predict exactly when this might happen.[6] Establishing the overall dynamics and structure of the system was far more important than precise outputs. The new approach allowed one to predict the future of the whole world in advance of actually having information about the whole world in the present. This meant that the idea of limits was very hard to refute with any particular set of data because the general principles of the model were all that mattered. But it was also hard to provide definitive evidence in its favor that would convince skeptics, who could just declare the basic assumptions about relationships to be wrong.

At the heart of *Limits to Growth* was an assumption that can be traced all the way back to Thomas Malthus's *Essay on the Principle of Population* of 1798: populations tend to grow in an exponential fashion. The report began by setting out the nature of exponential growth for the lay reader and then described how this pattern could be found in both population behavior and modern industrial production.[7] It then moved to examine the most basic resources needed to satisfy such growth, namely, agricultural land and the supply of nonrenewable materials (making estimates as to how long demand could be matched if reserves were, for example, five times larger than their currently known extent). Finally, the report assessed the resultant output of pollutants, beginning (already in 1972!) with carbon dioxide levels in the atmosphere but ranging to chemical inputs into lakes and seas or concentration of pesticides in human body fats.

The model then presented how all of these factors were interlinked by *feedback loops*; how nutrition was a function of agricultural output divided by the population; how levels of nutrition in turn might affect mortality and thus population levels, while levels of agricultural output also produced pollutants that affected mortality. Estimates were made as to the relationship between factors such as energy and raw materials demand and gross national product and equally how rising incomes would depress fertility and hence population growth. The real virtue of the model was not the predictions that various runs of the model yielded, which could only roughly approximate reality, but the potential for testing various scenarios. In a world of scarce raw materials, the model predicted "overshoot and collapse" would happen relatively soon. However, the assumption of unlimited energy supplied by nuclear power and more efficient technologies of extraction could sustain growth for much longer, until pollution caused catastrophic declines in life expectancy. Collapse could be postponed if stringent pollution controls were factored in, but then one hit a limit of land for food production instead. In the end there was no avoiding the concept that "the basic behaviour mode of the world system is exponential growth of population and capital, followed by collapse."[8] Technological advances could only delay the inevitable. The recognition of limits demanded a conscious decision to acknowledge them and seek a new global equilibrium.

What was new about *Limits to Growth* was the manner in which expertise was deployed to deal with what was actually a familiar problem. The fact that cutting-edge technology had been used to demonstrate the case gave it a certain frisson and legitimacy. As the American ex-secretary of the interior and leading environmental campaigner Stewart Udall put it, the model "made us respectable."[9] Equally, this was trust in numbers exhibited in a way never seen before and on a global scale.[10] No one locality could hope to evade the relentless logic of system dynamics. The understanding of feedback loops in turn shaped what counted as relevant information for modeling the environment, altering the substance of the idea itself. The environment of World3 was one of "quantitative restraints," understood as a numerically specified dwindling supply of raw materials and an

ever-growing mass of pollutants.[11] But most of the expressed concerns were not very distant from the issues of the "conservation" of
resources discussed at the United Nations Scientific Conference on
the Conservation and Utilization of Resources (UNSCCUR) at Lake
Success in 1949 (see chapter 2). Indeed, we could go further. The
Limits to Growth report might be considered an epitome of the style of
thinking that emerged in the postwar years, combined with the new
modeling and predictive technology that became available with the
computer. In the twenty-first century there is hardly anyone who
does not apply their methods as a matter of course. Equally, the questions they asked still lie at the heart of debates in economics and
environmental policy. Hence anxieties around resources and how
they became imagined and communicated are a starting point for a
more detailed examination of the areas of expertise that fed into
shaping the modern idea of the environmental, before we move to
histories of the understanding of ecology and climate.

Needs and Resources

In the wake of World War II, a blizzard of numbers blasted across the
desks of academics, civil servants, and policy experts of the developed world: census data and new methods for measuring national
income, such as gross national product;[12] figures on the consumption
and reserves of minerals and fuels; the accelerating needs of globe-
spanning military infrastructures; assessments of the impact and requirements of the Marshall Plan, a massive effort to reflate the European economy and sustain demand for American goods. New centers
for the collection and analysis of data sprang up, such as the Organisation for European Economic Co-operation from 1948, later known
as the Organisation for Economic Co-operation and Development
(OECD). Many had the imprimatur of the UN organization: UNESCO
in Paris, the Food and Agriculture Organization (FAO) in Rome, with
their staffs, borrowed experts, and steady footfall of committees,
conferences, and the solid, sober wisdom of "reports." The still dark
and hungry years of 1946 and 1947 gave way to the dawning of Europe's "golden age" of growth and the hope of a wider blossoming in

the economies of the "underdeveloped" world, but throughout this era, anxieties about resources surfaced again and again. Where would the resources for growth come from in a battered and impoverished world? The explosion of concern about resource scarcity was associated with the fastest rates of population and economic growth the world had ever seen. "We live in the hour of glory and of fear," wrote Osborn in *Limits of the Earth* (1954), his follow-up to *Our Plundered Planet*.[13] The future was inherently uncertain. As a motto for *Limits of the Earth* he had chosen these words: "To all who care about tomorrow."

In the postwar era, no country's demand for resources could touch that of the United States. With 10 percent of the "free world's" population, it consumed half of that world's resources, a disparity never matched in history before or since. It became commonplace in the 1950s and 1960s to compare the United States and India to illustrate the enormous disparity in their demand for goods and the consequence of India catching up with the Americans. Such an example was provided by Samuel Ordway's paper at the 1955 Princeton conference (see chapter 2).[14] In 1954, the population of the United States rose by three million people and that of India by five million people. If the iron consumption of each country was imagined to be equally distributed among all of its inhabitants, that new cohort of American babies, only at the very beginning of the "boom" in births that would last into the 1960s, would have accounted for 40 percent more iron consumption than all the 350 million people who lived in the subcontinent. American consumption per person was over one hundred times higher than Indian. Concerns about the "conservation" of resources became powerfully articulated in the corridors of American power. It was also an outcome of the war, because the system of national accounts applied to assess and harness "the war potential" provided the means by which predictions could be made about peace. This data was harnessed by the report *America's Needs and Resources*, commissioned by the progressive think tank the Twentieth Century Fund and published in May 1947.[15] There was no use of *environment* in the 812-page tome. Rather, it presented a markedly sanguine assessment of the nation's prospects: "we should not be hampered in meeting future needs by a shortage of raw materials."[16]

By 1951, on the back of a specific shortage of tungsten used in precision tools and munition making during the Korean War, the picture looked less rosy. President Truman established the President's Materials Policy Commission, better known as the Paley Commission.[17] This time the full resources of the government fell behind the collection of information, and the conclusions were less sanguine. Its final report was entitled *Resources for Freedom*. The "materials problem" was "pervasive," the question now being "has the United States of America the material means to sustain its civilization?" The answer, projected up until 1975, remained largely positive, but nevertheless a clear agenda was set that "upon our own generation lies the responsibility for passing on to the next generation the prospects of continued well-being" and the need to avoid "stagnation and decay."[18]

It was national governments or think tanks thinking about nations that set the pace for the analysis of potential postwar resource scarcity. What was striking to many Americans by midcentury was that the range of materials required for modern technological advance could no longer be easily supplied domestically even in their vast and varied land. The age of the frontier was closed; now it seemed to be shrinking. Recognition of the dangers of growth was thus simultaneously global, yet deeply nationalist. As the British left-leaning think tank Political and Economic Planning's report of 1955 stressed, "the limitations of considering these matters as if the world were all one place within which surpluses could be distributed, deficiencies made good, and populations moved from one point to another. . . . To that extent there is not a world population and resources problem, but a series of quite different national problems."[19] Yet, as the Paley Commission's conclusions forcefully argued, the global context and wider markets would shape the capacity of any one country to develop. Resources were a geopolitical issue. Access to resources was posed as a problem both for the continued development of the industrialized world and the possibility for what would soon be called

the "Third World" to develop at all—preferably, in the view of Western governments, without falling prey to communism. Both economic nationalists and advocates of free trade shared these ideas.
Whether one liked it or not, as Osborn concluded soon after, "Man
is becoming aware of the limits of the earth. The isolation of a nation, or even a tribe, is a condition of an age goné by."[20] On the basis
of the *Resources for Freedom* report and with support from the Ford
Foundation, in 1952 a research institute was established in Washington, DC, which would publish a series of landmark reports and
resource audits.[21]

The close association between national and global interests found
expression in concepts such as "peak oil," originally dubbed "peak
petroleum." The basic idea was simple. The most accessible petroleum reservoirs were being sucked dry, as already seemed to be happening in the United States. Costs of further expansion would be
high, and at some imminent point in time production might even
begin to fall. But demand was rising exponentially, and this could
only lead to steeply rising prices, especially after the "peak," when
supply began to reduce. On the other side of this cliff edge, if demand did not slacken, reserves could disappear very quickly indeed.
In fact, this wasn't America's first flap about oil; Cassandras had raised
fears of exhaustion after the surge of demand during the First World
War, which led to the government setting aside for the future the
protected region of the Naval Petroleum Reserve in Alaska in 1923.
Since then, ironically, it operates as a vast nature reserve although
drilling interests fight for access.[22]

The idea of "peak petroleum" was originally applied only to the
United States when raised in a 1949 article by petroleum geologist
M. King Hubbert, then working as an analyst for Royal Dutch Shell.
The article was discussed at UNSCCUR at Lake Success, where future energy supplies were a major issue.[23] Hubbert became famous
when he neatly illustrated the idea with a famous bell-curve graph
predicting in 1956 (very accurately, as it turned out, at least before
the expansion of shale oil) that American "peak oil" would be reached
around 1970. The argument over peak oil rumbled into the twenty-

first century, personified in the opposed positions and methods of Hubbert, who moved from his job at Shell to the US Geological Survey (USGS) and was a regular at environmental conferences in the 1960s, and his rival Vincent McKelvey, who was far more optimistic about the size of petroleum reserves and was appointed head of the USGS in 1971. Each reached to a range of technical and imaginative arguments to bolster their respective positions. McKelvey, in addition to using his power as Hubbert's boss to withdraw the latter's secretarial support, went so far as to develop his own theory about how technical ingenuity would overcome short-term resource limits. He formalized his argument with the equation $L = R \times E \times I/P$, where L was quality of life, R resources, E energy, I ingenuity, and P population. It was an equation whose operational uselessness Hubbert rightly mocked.[24] But the debate between optimists and pessimists over fossil fuel reserves proliferated in publications during the 1950s and 1960s. Notably, this was a process internal to the oil industry among analysts and geologists, rather than being initiated by external critiques of growth in general. Thus, the end of oil was a familiar trope in all the predictive literature on the state of the Earth from the 1950s.[25]

It wasn't just oil. Demands seemed to press on the limits of supply across nearly every resource: metals, water, energy, soils, forests, fisheries, whales. . . . In the late 1940s such thinking was aided by the rather loose and unsubstantiated assumption that it was precisely population and economic pressure on resources—with all that talk of *Lebensraum* and the Japanese resource grabs in Manchuria and Southeast Asia—that had led to World War II. If resource grabs caused war, then ensuring access to and wise use of resources could prevent it. "Conservation is a basis for permanent peace," as President Roosevelt put it in June 1944.[26] Yet the shared belief in a problem did not generate a consensus around a solution. Some authors, such as Fairfield Osborn and Samuel Ordway, stressed limits very explicitly. For them, growth was necessarily finite, and the answer was a turn away from a relentless pursuit of material gain to a new (or nostalgic) set of values. If growth was a problem, no growth was a solution. For others, this was a call to ingenuity, to better harness the resources that were available and invest in the technologies of the future. Nothing exemplified these trends better than the sometimes wild optimism surrounding nuclear power.[27] In Europe, the turmoil and bitterly cold winters of the postwar years, labor shortages after the war, and the fact that coal was being imported from across the Atlantic created fears about the adequacy of the coal industry and the promotion of a breakneck expansion of mining.[28]

Ironically, by 1958 Europe faced a coal glut. While tens of thousands were employed in unprofitable mines, cheaper oil beckoned.[29] Yet by the early 1970s, "energy crisis" was back on the international agenda, soon followed by the boycotts and blockades that began in October 1973 in response to the Yom Kippur war. It became a commonplace that these events were the warning signs of impending global oil shortages, and the memory of the "oil crisis" has largely superseded the idea of an "energy crisis" that was already widespread before 1973. Scoping possible energy futures and anticipating supply shocks became a minor industry in its own right, driven by business, national governments, and the new International Energy Agency (IEA), set up in 1974 to manage allocations of oil among members of the OECD.[30]

Like the problem of "the environment," the problem of resources became "scaled," in that every local problem was a subset of a planetary one. Simultaneously, the question of how to secure supplies of key resources became fundamentally oriented toward the future and what one believed about predictions of future supply and demand. Some observers treated every uptick in the price of a resource as an indicator of future problems. "Growth," as economic development came to be called at this very time, seemed desirable to eradicate poverty, to resist the allure of communism, and as the route toward a more amorphous "progress." Yet endless and especially exponential growth also seemed impossible on a limited planet. And nowhere did this seem truer than with an exponential curve rising faster than at any point in history: world population.

The Power of Population

It was hardly novel that there might be problems in the relationship between human wants and the capacity of the Earth to meet them. The idea is so associated with the writing of English clergyman Robert Thomas Malthus (1766–1834) that *Malthusianism* has become shorthand for this concept. Malthus did not so much see growth as a problem but as impossible in the *long run*—at least in the sense of the word used by modern economists who define growth as sustained rises in income per person. Malthus held that the total economy could be a lot *bigger*, but he argued that individuals living within it, in the long run, could not be any *richer*. He was more of an optimist than later prophets of doom who envisaged a world actually running out of resources; in fact, it is little known that he explicitly denied the likelihood of such an event. Malthus's great contribution was perhaps the strikingly clear way in which he formulated the issue of population pressing on resources in his classic *An Essay on the Principle of Population*, the first edition of which appeared in 1798.[31] It was intended as a rebuttal of what seemed to Malthus to be naïvely optimistic thinking associated with the French Revolution, particularly the work of Nicolas de Condorcet (who supported the Revolution but who died in prison during the Terror) and William Godwin (the English political philosopher, husband to feminist writer Mary Wollstonecraft, and father of author Mary Shelley). They saw the removal of the restrictions of the *ancien regime* and new policies for education and support for the indigent as routes to a permanently wealthier society. For Malthus any temporary benefits (and he strongly doubted these) would be undone by "the power of population . . . indefinitely greater than the power in the earth to produce subsistence for man."[32] This was *Limits to Growth* as played out by gentlemanly experts of the eighteenth century, although in that age of revolution, the political stakes were very high.

The predictive power of Malthusianism was focused on his idea of the *trajectory* that all societies must be on and where they must end up. It said nothing at all about *where* societies might *be* on that trajectory and hence when crisis or a steady state might be reached. This is not to say that Malthus was uninterested in the evidence he could obtain about the real world. He drew on the calculations of birth and death rates compiled by the German pastor and pioneer of demography Johann Peter Süssmilch, speculating that it was "highly probable, that a scantiness of room and food was one of the principal causes that occasioned the repeated occurrence" of spikes in the death rate identified by Süssmilch. Indeed, he labored the idea that his predictions were simply a reflection of a constant: "the period when the number of men surpass their means of subsistence, has long since arrived; and that this necessity oscillation, this constantly sub-sisting cause of periodical misery, has existed ever since we have had any histories of mankind, does exist at present, and will for ever con-tinue to exist."[33] This view can be contrasted with the open-ended optimism of Condorcet, who acknowledged the theoretical limits that the land might impose but shrugged them off: "It is equally im-possible to pronounce for or against the future realization of an event, which cannot take place, but at an æra, when the human race will have attained improvements, of which we can at present scarcely form a conception."[34]

The Malthusian framing of growth was rather different from the specter of resource exhaustion, but fear of things simply running out had a longer history too. Already in late medieval times we see the rise of fears about "wood shortage" in Europe, sometimes to an almost cacophonous extent from the wide forests of Sweden in the north to the wooden piles of Venice in the south. By the end of the sixteenth century nearly every European state had introduced legislation to protect wood reserves (enforcement, however, was another matter).[35] However, recurrent fears of "timber famine" were not treated as a reason for restraint but, rather, as a justification and stimulus for a more productive and well-managed forestry. By 1865, Frederick Starr of the US Department of Agriculture declared the United States was like a giantess who had "slept because the gnawing of want had not wakened her. She had plenty and to spare, but within thirty years she will be conscious that not only individual want is present, but that it comes to each from a permanent national *famine* of wood."[36] Clamour about timber famine and deforestation fed into broader fears about overexploitation of the Earth around the turn of the twentieth century.[37] In the case of timber we can see a merging of local fears, often based on expectation as much as evidence, into a worldwide prognosis of scarcity and attempts to create a global inventory of forest cover.[38]

As the world shifted its energy economy toward fossil fuels, similar logic was applied to this nonrenewable "subterranean forest."[39] In 1865, the very year that Starr imagined America as a giant sleeping its way to disaster, the British Parliament debated with great gusto fears of coal shortage fanned by the young political economist William Stanley Jevons's book *The Coal Question*.[40] The development of Britain's mineral statistics, censuses, and trade figures from the 1850s had provided a basis for calculating both the stock and possible future consumption of fuel. Jevons had worked as a gold assayer in Australia and as a meteorologist and was trained in mathematics and chemistry; he was an innovative and imaginative polymath. His work on coal recognized, although did not consistently apply, what would be a key insight for the important contributions he would make to economics before his untimely death in a swimming accident in 1882.[41] Jevons noted that the efficiency of coal use was likely to change with new technology, but, in turn, these savings would actually stimulate future economic growth and thus not actually reduce consumption. Bluntly, if you make something more efficient you are likely to both use it more and invest the savings in something else, so it's not clear that efficiency will lead to any less resources being consumed— an idea called the "Jevons paradox." Jevons's work also pointed to the way in which predictions would be made and tested in what were later called "feedback loops," an important part of the World3 model. However, the Royal Commission on Coal that had been appointed in the wake of Jevons's book concluded in 1871 that coal stocks would be plentiful for the foreseeable future. The excited estimate of Prime Minister William Gladstone in 1866 of the dizzyingly large size of demand for coal a century hence proved to be wildly off the mark by a factor of ten—as would be wryly noted in the report of Political and Economic Planning in 1955.[42]

The perceived desirability of taking inventories and predicting the availability of certain key resources was reinforced during World War I. The first global conflict brought together a wide range of expertise in the service of the state. After the war, population re-emerged as an urgent issue, a global problem that put pressure on agricultural capacities and required concerted action.[43] This was the idea behind the first world population conference held in Geneva in 1927, which included luminaries from economist John Maynard Keynes to birth control campaigner Margaret Sanger. This conference incorporated the Seventh International Neo-Malthusian and Birth Control Conference and hence had a strong orientation toward managing limits. The carrying capacities of the world's agricultural producers were contrasted with empirical predictions of population growth.[44] As census data improved over the twentieth century, the Malthusian debate could increasingly employ *actual* numbers. Advances in statistics delivered methods to develop predictive models, forerunners of those methods adopted by computer modelers in the second half of the century.

The Power of the Model

If the rise in fears of overpopulation and resource exhaustion was at least partly due to what we can call "real-world problems," nevertheless some of the inspiration came from a very different place: the laboratory. The American biologist Raymond Pearl, who was briefly a patron of Rachel Carson, promoted the logistic (S-shaped) curve as a standard and reliable geometrical technique to predict future human population developments within "definite limits"—including national ones. Pearl and Carson had worked with this in modeling fruit fly populations. Population growth across species was assumed to follow an S-shaped pattern of a slow initial advance, then exponential increase leading to saturation of the environment in which they lived, followed by collapse. Hence only a few data points could be used to predict future trajectories.[45] Another influence on midcentury population theory was Alexander Carr-Saunders, author of *The Population Problem* (1922), who trained as a zoologist under leading statistician Karl Pearson and went on to lead a famous Oxford University ecological expedition to Spitsbergen in 1921, where he worked on the book. Carr-Saunders's inspiration came from fieldwork studying the fluctuation of animal populations in isolated conditions. The expedition included among its members ecologists Charles Elton and Julian Huxley (see chapter 4). Carr-Saunders's volume became widely influential, but he also fostered the careers of protégés like Elton, who promoted modeling of animal populations and whose *Amimal Ecology* (1927) became a classic work on the topic.[46] Influence went in both directions. Ecologists were influenced in their understanding of the animal world by the work of demographers operating with Malthusian assumptions. At the same time, researchers on other species in the natural world gained insight into the population dynamics of humanity.[47] Sometimes this combination of interests and perspectives was driven by practical concerns. Raymond Pearl and another significant figure in this milieu, Harvard biologist and geneticist Edward Murray East, had been given administrative responsibility for securing food supplies during World War I and its aftermath. Pearl blamed population as one of the causes of the war itself, as did successors after World War II.[48]

Pearl also mentored the work of the extraordinarily influential mathematician Alfred J. Lotka. Born in Lviv in modern-day Ukraine, Lotka was raised in America but had an eclectic education in England, Germany, and America, working along the way as a chemist, editor, and mathematician.[49] Never securing a post in academia, he would eventually settle down as an insurance actuary in New York City, but his ravenous curiosity led him to develop influential mathematical approaches to a whole range of disciplines. He was himself an exemplar of the emergence of new forms of expertise, contributory and interactional, as his mind wandered across disciplines normally taught and practiced quite separately. Lotka sought to draw the varied problems of these disciplines into a common form of analysis. He generated statistics on subjects ranging from age distribution in populations to epidemiology to bibliometrics (the analysis of reading and book use).[50]

Lotka is best remembered for his research on predator-prey relations in ecology and energetic flows between species. He understood these as systems governed by feedback loops that could be expressed in abstract form in mathematics, an approach closely akin to the thinking of electrical engineers who also studied energy flows. Partly inspired by Herbert Spencer (see chapter 2), although by no means uncritically, he treated the systems of biology and the "industrial regime" as analogous to, or even bound into, an "organic whole" that in the end constituted a "World Engine."[51] Lotka's vision was a powerful inspiration to the Odum brothers, Howard and Eugene, who in the years after World War II became leading figures in presenting ecological relationships by directly lifting designs and symbols from diagrams of electrical circuit boards. The essence of ecological systems became conceptualized as flows of nutrients and energy. Eugene Odum's *Fundamentals of Ecology* (1953) became the standard textbook of the whole discipline for many years.[52] By this means the ideas of Lotka and 1920s thinkers became normalized in the training of ecologists: the world could be abstracted into fundamental flows of energy and matter treated in a way analogous to the dynamics of mechanical relationships or computer systems. Experts were expected to master these techniques. And they could be used to predict.

The linkages between statistics, ecology, demography, aspects of economics, and geopolitics were crucial precursors to the postwar understanding of the environment. The petroleum geologist Hubbert, for example, revisited the work of Lotka and Carr-Saunders in developing his predictions of peak oil from the late 1940s.[53] They provided the conceptual and technical infrastructure for developing modeling and the principle that the behavior of different things could nevertheless be captured by a common language—a language of meta-specialization and a new kind of interactional expertise.

Thus by the late 1940s Malthusian anxieties could draw on a long intellectual and technical heritage. The persistent worry about overpopulation was a concern that caused demographers, economists, agronomists, and others to come together. The distribution as well as the size of world population was commonly related to resource availability. Experts sought to predict growth rates and subsequent migration flows, often with an eye to geopolitical concerns that were frequently, but by no means always, entwined with racial preoccupations. The concern for "space" and population density drew analogies from animal ecology and for some created obvious links between questions of colonial expansion (most often related in the 1930s to the ambitions of Japan and Germany), conflict, and the dynamics of unchecked growth in human numbers.[54] In this milieu Vogt and Osborn developed their 1948 critiques of environmental destruction as an integrated global problem having roots in expanding human populations. Simultaneously, the logistics of global conflict had illustrated that certain finite resources were essential to military might and economic success.

The prewar preoccupation with space was gradually given a more explicitly ecological lexicon as it became a central aspect of a new postwar idea of environment. As an exemplar of this thinking, and indeed a very typical kind of integrative expertise, we may take Georg Borgström, a Swedish biochemist who turned into a Malthus in Michigan from 1956, having left his original discipline and found a new home in studying geography in America. He would return to Sweden as a respected environmental expert in the 1970s.[55] Borgström's chief concern was food shortage and food production, which he outlined in The Hungry Planet in Swedish in 1953, with an amended version appearing in English in 1965. He was relentless in his pursuit of issues such as topsoil erosion, waning soil quality, and the loss of virgin forests. Borgström was, if possible, even more pessimistic than Vogt and Osborn. He also brought a stronger apocalyptic element to his prose, talking of sin and punishment, of a cosmic doom that would come down on humanity should it not heed the calls of the new environmental experts. Borgström's hellfire and brimstone rhetoric betrayed his upbringing as the son of a pastor. He was not distracted by the talk of totalitarianism or loss of freedom that preoccupied his American peers.[56]

Borgström's message was apocalyptic, but his methods were nevertheless altogether quantitative. His books are loaded with tables and graphs. An increasing number of humans was the fundamental problem. New editions of his high-selling books, appearing almost annually from 1962, typically provided an update on how the world population had grown since the previous edition. As people overharvested the land and seas, Borgström sought to whip up fear. Many of his books visualized population by wedding a graph of exponential growth to the mushroom cloud of an atomic detonation. The authority of the predictive graph delivered by a technical expert was linked to the most striking imagery of humanity's self-annihilation. This was a double whammy of "interactional expertise." Humanity's long history of development, the stalk of the mushroom, explodes into the A-bomb cloud of population explosion in a brief moment of a few generations.

Yet contrary to the stereotypical Malthusian, Borgström was extremely positive about innovation and refused to be labeled as a pessimist.[57] In common with other researchers in the 1950s, he speculated that if humans could extract food lower in the trophic chain there would be ample resources available. One could exploit algae as a staple food or fashion a pudding out of grass—reminiscent of contemporary research that even posited sawdust as a source of nutrition.[58] In the late 1950s he began to talk of "ghost acreage," which described how countries' food demands were exceeding local carrying capacity and becoming dependent on land elsewhere: Holland, for example, had a large "ghost area" thanks to its food imports and active fishing. These ideas would experience a revival in the 1990s in discussions of the inequitable impact of development and what we now refer to as unequal "ecological footprints." Borgström pointed out that the strategy of dependency on ghost acres adopted by the developed world could not be repeated by every nation developing in their wake. Space, as the Malthusians had argued all along, imposed a fundamental limit. Indeed, the areas suitable for food production were actually shrinking because of growing deserts and rising sea levels, both at this point assumed to have anthropogenic causes.[59]

The most famous Malthusian work of the 1960s was Paul Ehrlich's *The Population Bomb*, published in 1968 and written rapidly at the behest of environmental campaigners at the Sierra Club, with a title clearly tapping into the anxieties of the age.[60] Ehrlich was a Stanford professor of biology and entomologist turned environmental seer. If possible, the pessimism had become even deeper: Ehrlich predicted at least ten million people starving to death each year in the 1970s and more thereafter. "The battle to feed all of humanity is over," he declared, and with a charismatic media presence he asserted the view frequently on national television and across print journalism, even in the unlikely pages of *Playboy* and *Penthouse*.[61] Thomas Robertson has already identified how much Ehrlich's arguments drew on that genre developed by Vogt and Osborn in the immediate postwar period. Massive US food aid to India in the mid-1960s had made prognostications of global famine a commonplace, but it is also the case that Ehrlich was emblematic of a particular kind of *expert*. His professional life as a theorist of evolution and lepidopterist (butterfly scientist) made him familiar with studying the physical limits found in ecology. But he also demonstrated interactional expertise, able, like Rachel Carson, to turn research into popular writing, deft in dealing with the public and author of punchy, pugilistic prose. He would test his writing on his twelve-year-old daughter for clarity (a habit that should perhaps be applied more widely).[62]

This new kind of expert had been in the making since the 1920s: the scientist who used his or her authority in a rather narrow specialism to step up and speak of the fate of the world. Other writers of this time who set contemporary challenges in a long sequence of humanity pressing against resource boundaries were the geochemist Harrison Brown, author of *The Challenge of Man's Future* in 1954 (see chapter 2), and biologist and professor of human ecology Garrett Hardin, who notoriously developed arguments against famine relief and in favor of strict immigration controls, an issue closely linked with arguments about overpopulation in the 1970s.[63] These writers thought explicitly at the level of the planet, and as the space race caught the imaginations of the 1960s, the metaphor of "spaceship

Earth" was famously employed to shape the imaginary of a generation of environmental campaigners, including architect Richard Buckminster Fuller and his famous *Operating Manual for Spaceship Earth*. The image was powerfully reinforced by the "blue marble" images of the jewel-like blue-green world hanging in the void beamed back from the Apollo missions.[64] Writers like Paul Ehrlich and Garrett Hardin even felt the need to present arguments as to the implausibility of colonizing other planets as a solution to overcrowding on Earth. Georg Borgström's book title of 1969 exemplifies many similar interventions by natural scientists into a new vigorous social thinking: *Too Many: A Biological Overview of the Earth's Limitations*.[65] The long history of Malthusian thought, and especially the techniques for integrating data and making predictions established in the 1920s, provided the context the computer whiz kids at MIT needed to model the dynamics of the whole world in *Limits to Growth*.

The Bet

Leading economists reacted critically to *Limits to Growth*. While an array of politicians, columnists, and scientists hailed the significance of the study, opponents accused its authors of making "arbitrary assumptions" and producing an analysis that was "worthless," no more than a naïve analysis by a "brash engineer."[66] Solly Zuckerman, the British scientist and policy advisor who had been instrumental in developing the environmental sciences and fostering integrative approaches, criticized "hysterical computerized gloom."[67] Critics included the authors' MIT colleague Robert Solow, who would win the Nobel Prize for Economics in 1987 for his contributions to theories of economic growth. The idea of a limit cut at the very heart of economics as practiced under the tutelage of Paul Samuelson at MIT, where the pursuit of growth had become the raison d'être of economic policy. In different parts of the same campus, methodological innovation was taking groups in very divergent directions.

This was not because the economists rejected the use of mathematization and modeling that gave such an aura to the *Limits to Growth* report. The contrary was the case. The rising mood in economics, typified by men like Samuelson, Solow, and Kenneth Arrow, was to use mathematical formulae to abstract the economy into very simple categories that could be imagined as a system tending toward equilibrium.[68] This looked a lot like influential trends in postwar ecology and population thinking. Samuelson had won his own Nobel Prize in 1970, cited by the Swedish Academy for doing "more than any other contemporary economist to raise the level of scientific analysis in economic theory"—which basically meant applying the rigors of mathematics. The methods that led to *Limits to Growth,* adapted from ecology, electronic engineering and systems theory, and the mainstream habits of postwar economics, had much in common.[69]

The development of gross national product as an indicator of total economic output, especially during the war years, provided a metric both of the success of the economy and a "fact" that needed to be "explained" by growth theory. The sum of rising inputs of labor and capital into economies could not, it seemed, by itself account for the quantifiable increase in the value of output in the American or world economy. In 1956 Solow identified the chief reason for economic growth as what came to be called the "Solow residual," the gap between the sum of labor and capital inputs, and the value of output: this accounting "residual" was assumed to be the quantified impact of technological advance.[70] Growth was thus not explained by rising resource inputs, the focus of Malthusians and theorists of ecological limits, but invention and ingenuity (indeed the fact of the efficiency of resource use increasing over time was already well appreciated by analysts and had been noted by Jevons in the formulation of his paradox).

Solow's riposte to *Limits to Growth* came at the prestigious Ely lecture to the American Economics Association in 1973. He argued that "the world has been exhausting its exhaustible resources since the first cave-man chipped his flint."[71] In fact, purveyors of the discourse of limits, such as Harrison Brown, had argued precisely this but noted that previous scarcities could be resolved by migration or expanding resource frontiers spatially or through innovation. At the planetary level, such solutions themselves became far more limited.[72]

However, for mainstream economists, this was not the point. One could not define what the essential resources of an economy were because they argued that history demonstrated people's capacity to find substitutes for what was running out. No economy would push itself over the cliff of limits, because as those limits were approached the increased costs of extraction or damage would prompt changes in behavior. This was "everyday market behaviour," quipped Solow. Critics of the World3 model noted that the abstracted population whose fate it modeled seemed incapable of reflection or the utilization of fresh information to change their circumstances, alter their reproductive behavior, or vary their tastes. In other words, the model did not reflect real-world conditions and markets.

These counterarguments were not novelties stimulated by the "limits" debates of 1972 but had developed over decades as economists became increasingly confident in the capacity of the market to signal potential problems and generate remedies. Scarcities would result in high prices and promote substitution of less scarce materials, while also providing incentives for innovation. Rising prices were temporary phenomena that said little about total resource availability. As oil analyst M. A. Adelman put it, responding to the "energy crisis" and renewed claims of peak oil in 1973, "The current scare (the nth) over the non-existent petroleum shortage hides the basic fact: companies explore not for oil but for cheap oil."[73] Actors in the market did not need to know anything about the complexities of ecology or the limits of the Earth. Prices would eventually compel rational actors to change their behavior to the net benefit of all and hence economic growth could be "sustained." A canonical text was *Scarcity and Growth: The Economics of Natural Resource Availability* (1963), by economists Harold Barnett and Chandler Morse, who essentially denied the possibility of limits in economic thinking. They argued that resources were made, not found.[74] What mattered was human resourcefulness.

While the Yale economist William Nordhaus mocked Forrester's team for producing "measurement without data," mainstream economics did not actually need any data to back its assertions because the assumption was that rational economic actors would change their behavior when hard times arrived.[75] This was not a wholesale argument against the possibility of environmental damage. Many critics accepted that the cost of environmental damage was not properly priced in the market, and hence policies such as environmental taxes could increase the incentive to reduce damaging behavior. A leading actor in this field continued to be Resources for the Future, the think tank set up after the Paley Commission. In 2010 it was awarded the Fondazione Eni Enrico Mattei prize for its work as "a key driver of market-based environmental policy."[76] In subsequent decades such approaches led to the paradoxical argument, building on the work of Harold Hotelling in the 1930s, that the environment was only destroyed when it was improperly valued by the market, and thus the solution was to create a market for "ecosystem services" (see chapter 7), even though this called into question the basic wisdom of the market in the first place. Of course, pricing the future value of a non-market resource might reasonably be seen as a different example of "measurement without data."[77] But the crucial point is that under the assumptions current in economics, there were no limits.

Thus the discourse of resource limits and the emergence of a more mainstream environmental economics confident that resources were only ever temporarily scarce emerged and developed in parallel—and parallel means never meeting. In a decade of "energy crisis" alarm about peak oil, the necessity of developing alternatives won support in government and even among some in the major oil companies. The United States launched its "Project Independence," indicating how economic nationalism and fear of dependency played a major role in resource anxieties.[78] The high point of the influence of such fears came with the Carter administration's *Global 2000 Report*, completed in 1978 but not published until 1980.[79]

A direct response to the *Global 2000 Report* was *The Resourceful Earth* of 1983, edited by conservative economist Julian Simon and futurologist Herman Kahn and fêted by the new Republican incumbents in the White House.[80] Simon had worked on population questions over the previous decade, increasingly irritated and dismayed by what he felt were overhyped environmental fears and poorly evidenced Malthusian ideas. He first clashed in print with Ehrlich, author of *The Population Bomb*, in a 1980 article in *Science*. Paul Sabin's *The Bet* vividly describes the duel between the two. Ehrlich had a habit of proclaiming "If I were a gambler" as he delivered prognostications of disaster, and Julian Simon decided to take him at his word. Simon offered him a bet on whether the price of key resources would increase or decrease over the 1980s. If Ehrlich's prognostications were right and overpopulation was eating up the last of the Earth's resources, then they should rise in price. Simon reasoned that this would turn out not to be true, or people would find adequate substitutes, and so prices would stay stable or decline. The final payout was to be determined by the size of the rise or fall. To make Ehrlich even more likely to bite, Simon offered that his rival could choose the commodities on which to base the wager. Ehrlich chose five metals extensively used in industry: chromium, tin, tungsten, nickel, and copper, the latter a metal that Simon predicted in the future could be made from other materials so that the final limits on its use were no less than the mass of the Earth itself![81]

It was "the scholarly wager of the decade." The basic assumptions of Malthusian doomsayers and techno-optimists were to be put to the test. Ten years later, in October 1990, Simon went out to collect his mail at his suburban house in Maryland. Inside the box, he found an envelope containing nothing but a list of metal prices and a cheque for $576.07 posted from Palo Alto, California—the home of Paul Ehrlich. The prices of all the metals had fallen, in the case of tin and tungsten by more than half.

Simon had won, and his triumph seemed symbolic of the rise and dominance of free market ideology and the economics mainstream (although this covers a wide continuum of opinions in which Simon stood on the libertarian right). The repeated grim prophesying from some leaders of the environmental movement seemed to many to be a case of crying wolf (especially to those who were unsympathetic in the first place), and "the bet" was further proof. However, the bitter dispute was, of course, not resolved. After all, the bet only covered a small time period and a subset of resources among what both men believed were unimpeachable wider dynamics. Some scholars have noted since that in other ten-year periods prices of the metals have risen, so Simon maybe just got lucky (although the bet was made in 1980 and according to the conditions prevailing then). While the optimists could take the view that this proved Malthusians wrong, one could equally argue that other, short-run effects in the economy temporarily altered the longer-term trend. Or maybe Ehrlich had chosen the wrong commodities? Julian Simon continued to decry what he saw as environmentalism's propensity to "ignore the scientific literature," leaving as he saw it "truth" under siege from the use of poor evidence and repetition of Malthusian predictions that failed again and again.[82] The ecologist Ehrlich, in common with many of his scientific peers and the pioneers of a new "ecological economics," insisted that the economy was still subject to the dynamics of the Earth.[83] For them, the proposition that one could endlessly invent and substitute one's way out of limits was absurd.

The techno-optimists and cornucopians were focused on how people derived value from resources, which to them seemed almost infinitely malleable. Scarcity was ever present but ever temporary. In contrast, Ehrlich and others saw resources as a finite part of a global system based on material flows they already knew to be limited. Resource anxieties were a powerful part of the emerging narrative of environmental threat because they were so widely shared. But that they played such a prominent role in the formulation of the environment, when mainstream theorists of resource economics tended toward optimism, was because of the integration of resource anxieties into ideas developed in ecology. It was ecology that inspired much of the neo-Malthusian thinking between the 1920s and 1960s, and it is to this discipline and its reimagining of the older tradition of conservation which we now turn.

Ecology on the March

Conservation Meets Ecology

The Dust Bowl of the American Midwest in the "dirty '30s" was more than an environmental disaster. It was a social disaster. John Steinbeck captured the mood in his famous 1939 novel *The Grapes of Wrath*. The "Okies," the small-block farmers from Oklahoma's panhandle who watched their dreams blow away with the soil, escaped westward: "The causes are a hunger in a stomach, multiplied a million times; a hunger in a single soul, hunger for joy and some security, multiplied a million times; muscles and mind aching to grow, to work, to create, multiplied a million times."[1]

As the midwesterners ran in panic, they made people farther west grow nervous, too, "as horses before a thunderstorm," Steinbeck wrote. The stampede of westering people, fleeing from all they had known and worked for, was a sign of trouble. Black clouds of blowing topsoil reached the cities, with the Okies in their wake. Soil erosion was the disaster. Conservation was its cure: with it would come progress, perhaps even civilization.

Australia, too, suffered a dust bowl in the 1930s. There, too, city skies darkened with storms of topsoil and people ran from the land, ashamed, in the night. The Oxford-trained ecologist Francis Ratcliffe, who traveled to inland Australia in the years of "drifting sand," was moved by the plight of the long-suffering farming families in the impossible climate: "The essential features of white pastoral settlement —a stable home, a circumscribed area of land, and a flock or herd maintained on the land year-in and year-out—are a heritage of life in the reliable kindly climate of Europe. In the drought-risky semi-desert Australian inland they tend to make settlement self-destructive."[2]

Was Western civilization only possible in "reliable climates"? Such questions made settler nations anxious.[3] The political leaders who in the 1920s promoted the expanding interior wheat and pastoral fields as "lands of opportunity" were later confronted with the moral challenge of soil erosion. To whom should they turn?

Ecologist Paul Sears in his great polemic against soil erosion, *Deserts on the March*, recommended a new expertise, "a point of view, which peculiarly implies all that is meant by conservation, and much more. . . . It is the science of perspective. . . . It is the approach to biological knowledge, which is called ecology."[4] For Sears, conservation was a way of life. Ecology provided essential technical knowledge to manage land, soils, and agriculture, but conservation was more than this: it included society and its aspirations for the future. *Deserts on the March* was part of a mission to impart the principles of conservation to schoolchildren. Already in 1935 the book's first edition had inspired Congress to establish an authority, the US Soil Conservation Service.[5] The second edition (1949) was for the public. Conservation "of our resources is not a subject. It is a moral attitude," Sears urged. He pointed out that one cannot teach attitudes, but one can offer leadership through "a science of perspective and holism."[6] Ecology thus became a tool to manage the future.

Conservation of land, of soil, and of wildlife was about partnerships and connections. It called for expertise on the ground at a local level, a "trained ecologist in each community." Sears did not write of the environment. For him, nature was a resource for humans, and it required stewardship: humans had a moral responsibility to make wise use of nature. Sears hailed from a generation that had lived through the Great Depression and had become concerned that modernity had overstepped limits. He drew on ideas and experiences that ranged from field science and soil management to international and imperial politics. Ecological thinking linked the destinies of nature and society. Ecology offered the technical expertise to rein in

literary education to move the hearts and minds of the Australian public and its agricultural policy makers with her *Soil and Civilization* (1946), independently following the same trope at nearly the same time, half a world away. Mitchell took the position that if "the fundamental history of civilization is the history of the soil" . . . then "civilization as we know it—art and literature; music, poetry and philosophy; cathedrals, houses, farms, universities and theatres—will go towards a rapid destruction unless we ourselves awaken to retrieve the land."[12] Unconsciously echoing Sears, and predating Vogt, Mitchell used parables from the Old World to reframe the New: "Age-old Egyptian cities are filled with sand. The Roman Empire made the deserts of North Africa. Mongols flooded Europe when deserts started to encroach on their pastures. Are we unconscious of history or deeply careless of the future? Pioneers in the New World, finding the richness of the soil in which for aeons of time nature had preserved harmonious balance, were without the experience to understand that, by not preserving the balance, they sinned against the future."[13]

The idea that soil was crucial to the health of the land and to future civilization itself was in the air, quite literally. "The dust-clouds are carrying with them the material that should be taking the shapes and forms of life," wrote Mitchell. Fearing that civilization and life itself would be "gone under the tide of man-made deserts," she appealed to "retrieve the land" through intelligent technical solutions and a change in attitudes. "There is a profound difference between a strong nation fighting to retain the proper balance of its soil in a poor, hard land and one rapaciously using up its soil's entire stock of vitality."[14]

Yet land is more than just soil, a point forester and animal ecologist Aldo Leopold made eloquently in his influential 1948 essay, "The Land Ethic." An overemphasis on national economy may be harmful to the biotic community: "Of the 22,000 higher plants and animals native to Wisconsin, it is doubtful whether more than 5 per cent can be sold, fed, eaten, or otherwise put to economic use. Yet these creatures are members of the biotic community, and if . . . its stability depends on its integrity, they are entitled to continuance," Leopold argued.[15]

Britain's Colonial Office supported government-funded science in Australia, New Zealand, and other dominions and colonies for agricultural developments in the 1920s, to "feed the world" at a time of anxiety about growing populations and the capacity of fertile regions to feed them.[16] Nationalism and patriotism were challenged by a new "world-mindedness," fostered by an awareness of the responsibility to growing populations, not just within the Empire but also beyond. Questions of desertification were also international problems, not just national ones. In 1951, India and Israel urged the United Nations Educational, Scientific and Cultural Organization (UNESCO) to establish an international and intergovernmental scientific program, the Advisory Committee on Arid Zone Research. Its focus was anthropogenic desertification—deserts created by human practices, carrying topsoil away with the winds. Growing food was becoming a (civilized) national obligation to contribute to global humanity, an aspiration continued in the rhetoric of "food security" in the twenty-first century. Future food security depends not only on self-sufficiency but also on trade. National concerns and global ramifications are interdependent. No nation could escape the context of global resource pressures, as discussed in chapter 3. This raised equity issues about demand and supply, highlighted in Georg Borgström's 1953 concept of "ghost acres," the places sacrificed to grow food for faraway cities.[17] Global projects for feeding the world through cooperative agriculture, research, and new technologies inspired the original International Institute of Agriculture, established in Rome in 1905, which in 1945 at a meeting in Quebec, Canada, became the UN Food and Agriculture Organization (FAO).

The new postwar civilization was built on soil and intensified its use, with an increased imperative to grow more food and fiber. By 1968, the "green revolution" in agricultural research, development, and technology transfer initiatives emerged, with a very different emphasis and expertise from the 1930s eugenicists and others campaigning for reducing population.[18] Whether the emphasis fell on conserving soil or expanding production, in the mid-twentieth century ecology was viewed by many of its practitioners as an applied discipline and one that might even decide the future progress of society—and the world as a whole. These landed concerns had their counterpart in the world's oceans, where both overfishing and the extinction of cetaceans were hotly debated, at the same time as desires were expressed to efficiently harvest the huge bounty of the seas.[19]

From Natural History to Ecology

Applied ecology had deep intellectual roots, stretching back to early modern inquiries in natural history that had predominantly been the domain of gentlemanly hobbyists rather than professional scientists. In the years between 1838, when the polymath William Whewell coined the term *scientist*, and 1901 when the first Nobel Prizes were awarded in the professionalized sciences of physics, chemistry, and physiology or medicine, natural history evolved from hobby to science. Among the many new lines of inquiry and associated neologisms of this age, the German Ernst Haeckel, a strong supporter of the theories of Charles Darwin, proposed the term *Oecologie* in 1866 as "the science of the relations of living organisms to the external world, their habitat, customs, energies, parasites, etc."[20] It took many years, however, to develop as a science, like Herbert Spencer's word *environment* from the 1850s. Ecology and environment were projects for the future. Even in 1893, when *ecology* developed its anglicized spelling at an International Botanical Congress in Madison, Wisconsin, it was regarded as a new subspecialty of botany rather than an independent science, a discipline to which one made contributions in its own right.[21] In this period, the term *ecology* was used only by a small group of professional botanists. As late as 1902, a reader of *Science* wrote to the journal asking the meaning of this new and obscure term.[22]

Unlike the worlds of physics and chemistry, scientists studying living things worked in fragmented disciplines: in botany and zoology; in classification, anatomy, and physiology; in medicine and in agriculture. The study of ecology offered a new avenue in that it described itself not by the kind of organism studied but rather by the relations between life and its abiotic support systems. It was both about life and about physical environments, but these were interdependent: there was no separate environment conceived, simply physiological responses to the physical properties of the surroundings.

Many more of the early ecologists dealt with plants rather than animals before the 1920s. Because it does not roam like an animal, a plant's surroundings were easier to define and measure. Climate was an important factor.[23] Growing food resources in different climates was also a project of agricultural science, and in the United States, many early ecologists were applied scientists, training farmers in land-grant colleges to get the most out of what were still relatively newly cultivated lands in the Midwest. As with the later experience of dust bowls, ecological expertise was closely related, in these regions, to the experience of the agricultural and settlement frontiers in areas relatively recently colonized by Europeans, and became subject to new demands and commercial pressures.

In Europe, a more theoretical and autonomous ecology emerged at the end of the nineteenth century. A leading figure was Eugen Warming (1841–1924), University of Copenhagen professor of botany. Warming was a polymath. He served on the board of the Danish Geological Survey and developed his interest in biogeography through travels in tropical countries, especially Brazil. Warming's classic ecological text of 1895, *Plantesamfund* (literally, "societies of plants"), immediately translated from the Danish into German in 1896, was based on the lectures prepared for his undergraduate ecology course, acclaimed as the first course in ecology in the world.[24] *Plantesamfund* set out "principles of ecological thinking" in a format accessible to students. The title prefigured later European interest in "plant sociology." It echoed the ferment of ideas influenced by Herbert Spencer that confidently drew analogies between society and the natural world and vice versa (see chapter 2).

By the early twentieth century, specialist ecological societies had been created: the British Ecological Society (BES) and the Ecological Society of America (ESA), which produced their own journals, *Journal of Ecology* (established 1913) and *Ecology* (from January 1920), respectively.[25] From the outset, ecologists sought to be useful, and the scale and conceptualization of their studies reflected this strong orientation toward applied science. The locations in which they worked and the transformations were important not just to their broad interest in ecology and for securing the funds for their studies but also to the focus of that research and the way in which they understood the natural world to work. As the prairies of the Midwest of America were coming under the plough, the natural succession of vegetation became the object of study. Ecology here focused on the "climax" or supposedly stable state in vegetational succession, the "natural state," which was being swept away by advancing farms. The vegetational associations of wild prairie plants became the basis for the theoretical work of Henry Chandler Cowles (1869–1939)[26] on ecological succession in the dunes around Lake Michigan; on bog-lake to forest succession in Minnesota by Raymond Lindeman (1915–42);[27] and the influential theories of Frederic Clements (1874–1945) on vegetational succession on the Nebraskan prairies.[28]

These all became influential far beyond the United States as models for a more general understanding of ecological change and stability. While Lindeman was interested in the trophic dynamics (the exchange of nutrients between different kinds of species) of the bog-lake system, Cowles and Clements theorized the natural history idea of the "balance of nature" in their regional context. They translated it into scientific terms, operationalizing it through experimental plots. All were passionate field workers, and the scale of observations in the field strongly influenced their view of ecology. Clements wrote of plant "communities" and associations, treating them as "superorganisms," thereby combining organisms, the relations between them, and their physical environment in a bigger entity of biotic and abiotic parts working as one. Amid the experience of agricultural transformation, these ecologists hypothesized a baseline of stability to which local environments might tend to progress if left undisturbed.

While most ecologists studied vegetation communities, Victor Shelford (1877–1968) and Charles Elton (1900–1991) focused on animals and conservation. Shelford's *Animal Communities in Temperate America* (1913) was a major work of both ecology and physiology.[29] Oxford ecologist Charles Elton was author of *Animal Ecology* (1927), the major textbook in ecology for zoologists for decades, revised most recently in 2001.[30] At a time when zoology generally focused more on physiology and evolution, *Animal Ecology* claimed "to offer immediate practical help to mankind." Elton's preface to his book extolled the civilizing powers of scientific knowledge in the imperial cause of agriculture: "In the present rather parlous state of civilisation, it would seem particularly important to include it in the training of young zoologists." Elton explicitly drew on the contributions of "people working on economic problems, many of whom were not trained as professional zoologists." He was strongly connected with scholars, researching demography and concerned with issues of overpopulation, who themselves drew insights from animal ecology (see chapter 3).[31]

Shelford was elected the foundation president of the ESA in 1915, but his efforts to steer the ESA into lobbying for practical outcomes for conservation fell on deaf ears when the mood within the society was to establish itself as a professional science group.[32] Elton's work, particularly his 1920s fieldwork in the Arctic, caught the attention of businessman Copley Amory, who was concerned about the ailing North American salmon fisheries in the Gulf of Saint Lawrence. Through applications for business, Elton made animal ecology an applied and management-oriented science, unlike plant ecology, which remained more theoretical and academic. Amory sponsored a major international biological conference at Matamek, Quebec, Canada, in 1931 and invited Elton to be its secretary. This event, with its practical conservation focus, in turn inspired the Oxford Bureau of Animal Population (BAP), established at Elton's home university in 1932. Many of Elton's students found themselves working abroad, "up against practical problems in the field."[33] This was the era of ecology for empire: the British Colonial Office employed more biologists outside Britain than worked within the country.[34] BAP graduates were highly sought after throughout the British colonies and dominions in Africa, Asia, the Caribbean, and the Pacific, often dealing with the ecology of pest species and vectors of disease. It was from the BAP that Ratcliffe, whom we met above, came to Australia in 1929 to work on the ecology of flying foxes (fruit bats, *Pteropus poliocephalus*) with the Council for Scientific and Industrial Research (CSIR) and returned to address the question of soil drift in the 1930s. He then stayed on to head its first Wildlife Ecology Division.[35] Ratcliffe was one of the rare individuals who worked in all the ecologies: in applied agriculture, in soil conservation, and in wildlife biology and conservation, weaving the disciplinary insights into different manifestations of "conservation," before and after the idea of the environment took shape in the 1940s.

New Ecology and the Ecosystem

The ecosystem has been one of the most crucial concepts for widening the influence of ecologists as environmental experts. Familiar to schoolchildren in many parts of the world, the ecosystem concept was part of a mid-twentieth-century trend for cybernetics and "systems" thinking. It helped make ecological ideas modern, scalable, and predictive.[36] *Ecosystem* was originally coined in 1928 by A. R. Clapham, a crop physiologist at Rothamsted, an agricultural research and experimental station just north of London. Clapham's background in physiology and physics oriented him toward an interest in systems theories and the development of statistical techniques based on analogies from thermodynamics.[37] But it was Arthur Tansley (1871–1955), professor of botany at Cambridge, who provided a functional definition for ecologists and has become closely associated with the concept. His 1935 paper made ecosystem the most integrative concept in his discipline in the 1930s and, arguably, since.[38] Despite more than twenty years of close collaborations between British and North American ecologists,[39] there was a distinct difference of emphasis between Clements's and Cowles's prairie-driven ecological theories of North America and that of Tansley and the continental Europeans, including Swiss ecologist Josias Braun-Blanquet, who had developed mathematical methods for comparing plant communities at the University of Montpellier in France.[40] Tansley also challenged the vegetational concepts used by South African philosopher-biologist

John Phillips, who further developed Clements's theories in a suite of papers in 1934 and 1935, "Succession, Development, the Climax and the Complex Organism: An Analysis of Concepts."[41] Phillips's work drew on the idea of "holism," coined by South Africa's famous and controversial international statesman Jan Smuts (also admired by Paul Sears).[42] Smuts's desire for seeing interconnection proposed the idea of botanies and ecologies that fitted political units, in the style of linking soil and civilization, but also incorporated people into such "naturalized" thinking in keeping with his place in racially segregated South Africa. He argued that upland regions were in fact best suited to European settlement and colonization, while African populations should be restricted to low-lying regions, which also fit their purportedly lesser level of civilization.[43] In the definitions of Phillips, however, the universe itself became an organism. Tansley thought these ideas too broad to be of practical use in a science designed to include modernity and physicists.[44]

Tansley's ecosystem paper, with the unlikely title "The Use and Abuse of Vegetational Concepts and Terms," was about conceptual definitions. Its primary aim was to keep ecology internationally coherent. A gap had emerged between ecology in the New World, where nature was treated as separate from the human landscape, and the Old World, where all landscapes were already cultural. Tansley argued for integrating the human into understanding landscape processes but resisted anthropomorphizing the systems themselves. "We cannot confine ourselves to the so-called 'natural' entities and ignore the processes and expressions of vegetation now so abundantly provided us by the activities of man. Such a course is not scientifically sound, because scientific analysis must penetrate beneath the forms of the 'natural' entities, and it is not practically useful because ecology must be applied to conditions brought about by human activity."[45]

Thus, Tansley understood ecosystems and their successional processes generally to include anthropogenic change. He excluded catastrophic events, whether a blundering elephant on a small scale or a volcanic eruption on the large, a view later ecologists would come to revise. Catastrophes "are unrelated to the causes of vegetational changes," he argued.[46] Yet the role attributed to humans overturned the idea that climate was the sole determining factor of climax (as Clements had advocated). Tansley considered an ecosystem to include a range of other drivers, particularly soil factors such as chemistry, texture, and capacity to hold moisture.[47]

Tansley's idealized scientific concept was designed to be scalable and relevant to the working situations where scientists found themselves. If ecologists were to become "experts" for working landscapes (and empire), they needed theoretical concepts inclusive of applied situations. Soils and climates were important elements in an ecosystem, "the habitat factors in the widest sense." More than a biome, an ecosystem became properly the subject for study by physicists, soil scientists, and chemists, along with biologists. Tansley thereby reconfigured ecology itself as a new meta-discipline or, as we might say, a meta-specialization, rather than a mere subdiscipline of biology. This step made ecology's central role in shaping expertise for the broader idea of the environment after the war all the easier. Increasingly habituated to a wider scale of thinking that included both biotic and abiotic factors, ecologists were the natural leaders in *environmental science*, when that term emerged in the 1960s.

The ecosystem was the signature of ecology's new modern, scientific edginess. It tackled the issue of poorly defined metaphors in the science and provided a new term that had traction and respect in the physical sciences as well as the life sciences. Systems theory (or cybernetics, as it would later become known) was evident all around Cambridge in the interwar years. While a suite of key ideas about the mathematics of networks was developing across many sciences,[48] ecology was straining to contain the understanding of key vegetational concepts. Tansley defined the *ecosystem* in a way that ensured a fundamental concept in ecology meant the same thing on both sides of the Atlantic—and therefore for "new ecology" throughout the world—and the term's use of *system* signaled connections with systems theory.

There were other ideas in the air as well. Julian Huxley, later a leading figure in the foundation of UNESCO (see chapter 6), defined ecology in 1931 as the study of the "balances and mutual pressures of species living in the same habitat" in his influential and popular *The Science of Life*.[49] Tansley, Huxley, and Elton were also all interested in the mathematical models of predator-prey relations developed by Alfred Lotka (see chapter 3) and Vito Volterra in the 1920s.[50] Ecologists became part of the intellectual shift toward modeling that we have already seen in population biology, demography, and resource economics and that became touchstones for the environmental thought of the postwar era. Tansley's intervention was thus part of a much wider mood that was building toward reconfiguring the way a range of disciplines conceived of the world. Like Elton and Huxley, Tansley was conscious that humans were just one animal among many and that human communities, wild animal communities, and domesticated animal communities had all changed landscapes in ways that concepts like the ecosystem needed to accommodate, not ignore. Tansley's theoretical purview knitted soil conservation back into the fabric of ecology just as dust bowls became a major political and practical issue.[51]

Nature Out of Balance beyond Agriculture

By the 1940s, ecology and ecologists were well set to provide a major contribution to the new discourse about the environment. Significant figures within the discipline had participated in the interwar movement toward cybernetics, modeling, and the study of numerical data. These trends would provide the underpinnings to a trend toward understanding ecological relationships in terms of general phenomena like energetics and, eventually, to the "big ecology" that could be integrated into the study of global systems change.[52] It was apparent that plants and animals (deliberately or accidentally introduced) could make massive ecological transformations. Sometimes large-scale transfers were only effective when constantly managed (as in agricultural crops), while small-scale transfers (for example, a handful of rabbits introduced to one farm in Australia in 1859) could have continental effects. When viewed on the scale above the traditional small experimental plots, dynamism might be more common than ecologists had appreciated, and it worked differently depending on the scale and context of the experiment.

Botanist John S. Turner (1908–91) arrived in Australia in the last days of 1938. A graduate of the University of Cambridge, Turner was appointed to the University of Melbourne's Chair of Botany and Plant Physiology before his thirtieth birthday. Before he had been in Melbourne a month, the Black Friday fires of January 1939 burned three-quarters of the state of Victoria (twenty thousand square kilometers), killing seventy-one people and destroying thirty-seven hundred buildings. In the face of these horrific wildfires, Turner realized the importance and urgency of ecology. The plants were different and so were the fires, in the dry continent of Australia, whipped by hot summer winds from the desert.[53] These were fires on a scale unseen before by Western ecologists. They directly challenged Tansley's formulation that catastrophes were unrelated to the causes of vegetational changes.[54]

In hindsight, in an interview in 1991, Turner "regretted the fact that he did not have sufficient background to seize that opportunity to study forest fire ecology."[55] He did, however, support new research in the postwar era, particularly that by ecologist David Ashton (1927–2005). The well-named Ashton spent a lifetime documenting the mountain ash (*Eucalyptus regnans*) forests near Melbourne as they regrew after the 1939 fires. Ashton's work revealed that such forests needed massive fires—indeed, the destruction of the whole forest— to release seed from the canopy and regenerate in ash beds in full sunlight.[56] Turner the Cambridge scholar confronted the stark need for a new ecological understanding in a continent with very different seasons, vegetation, animals, and even ecological drivers from those of North America and Europe, where the key concepts of ecology were developed.

Aboriginal people had used "fire-stick farming" (deliberately lit cool fires) to manage the country for over sixty thousand years.[57] Before this, the Australian vegetation had become fire-adapted in response to lightning strikes and drying out over millennia. The dominant and distinctive *Eucalyptus* and *Acacia* vegetation is well adapted to survive in the poor soils and dry conditions of the continent. Aboriginal people had learned to tame the natural wildfire through cool burning to reduce fuel loads early in the season. Burning was good for green "pick," attracting animals to hunt.

Thus ecologists discovered completely new theoretical and practical insights as they worked in new places. They became increasingly aware that plants and animals functioned differently in different ecosystems. The long-isolated flora of Australia behaved differently when transplanted to Africa (*Acacia*) and California (*Eucalyptus*).[58] Trees that were moved so they would be economically useful became problematic, extracting limited water supplies in southern Africa and causing new, hotter wildfires in places like California and Portugal. Animal introductions such as rabbits and sheep often created havoc for native species and soil stability. Accidental introductions (such as rats escaping from shipwrecks) also contributed to the global spread of "invasive species." Human diseases had also been part of the "Columbian Exchange," causing mass deaths of peoples without a genetic history of exposure.[59] In the postwar years, the ideas of global ecological invasions and the consequences for the planet of these widespread transfers became a problem demanding ecological expertise. *The Ecology of Invasions* emerged from a BBC radio broadcast, "The Invaders," delivered in 1957 by Charles Elton, now an elder statesman of science. He reflected on what it meant to "conserve" the animal kingdom in a global world, in a program entitled "Balance and Barrier." Invasive animals upset the balance of nature (still a key concept) as they crossed the natural barriers of the world (oceans and deserts). The "barrier" hinted at the military territorialism possible through quarantine stations and picked up on the strategic as well as the natural "barriers" that might be managed through invasion ecology. In another broadcast, Elton examined animal distributions through the work of nineteenth-century natural historian and biogeographer Alfred Russel Wallace, perhaps best known as Darwin's co-discoverer of the principle of natural selection. Elton argued that "if we are to understand what is likely to happen to the ecological balance in the world, we need to examine the past as well as the future."[60]

The problem of balance and imbalance in nature was epitomized in the case of "biological invasion," especially in places beyond the Old World. Invasion was also something that happened at many scales and could be studied in any stream or pond, as well as in working rangelands and deserts. Invasions transformed environments on regional and even continental scales, so invasion biologists worked at all scales but often found it difficult to consolidate conclusions across these scales. The Scientific Committee on Problems of the Environment (SCOPE) program launched an international assessment of the ecology of biological invasions in the 1980s.[61] The scientific questions of the 1950s—and the explosive nature of invasions identified by Elton—continued to drive research in invasion biology, a prominent practical subdiscipline of ecology. SCOPE asked (1) What biological characteristics make an invader? (2) What makes a natural ecosystem susceptible to invasion? (3) How can science predict (quantitatively) the outcome of any introduction? (4) What is "best practice" for managing and conserving natural and seminatural ecosystems? These questions framed scientific practice from the local to the international. Ecology had geopolitical implications. Elton's imaginative leap of conceptualizing biota as *invaders* opened an expert niche for ecologists waging a "fight against invasion," which appealed to the scientists that followed him. Half a century later, Elton continues to be important to the professional identity of invasion biologists.[62]

Yet at the same time, management applications generated tensions for ecological theorizing. Practical issues such as wildfire and invasive species exposed dynamic, nonreversible changes. Despite Tansley's carefully dispassionate language when he described an ecosystem, the attraction of "balance" was always there, and Elton's notion of "invasion" clearly played upon this. Ecologist and author of *Discordant Harmonies* Daniel Botkin remarked: "If you ask an ecologist if nature never changes, he will almost always say no. But if you ask the same ecologist to design a policy, it is almost always a balance of nature policy."[63]

Even Eugene and Howard Odum's important 1953 *Fundamentals of Ecology*, the leading ecology textbook for many decades, used the metaphor of balance: "Living organisms and their nonliving (abiotic) environment are inseparably interrelated and interact upon each other," they wrote. "Any entity or natural unit that includes living and nonliving parts interacting to produce *a stable system* in which the exchange of materials between the living and nonliving parts follows circular paths is an ecological system or ecosystem."[64]

Life "cycles" from birth to death to decay to rebirth in all the volume's diagrams. Yet the ecosystem includes the abiotic too—and energy from the sun is an input, not a cycling factor. In the Odum textbook energy systems are important, especially the way plants capture energy from the sun, animals capture it from plants, and so on. But the life sciences focus on "life" cycles, and the environment is more than life—it is all the different factors that support life. Life scientists needed physics and chemistry as well. A truly dynamic Earth system is much more than the pond or the lake or other small-scale life systems that were typical of the traditions of nineteenth-century natural history. Ecology provided a set of concepts that portrayed a world spinning out of control through population growth and transformations to the environment in the postwar period. At the same time, ecology met its own limits in describing the Earth as a system and understanding the dynamic interplay of its parts—whether led by human action or not. This would lead to ecology "scaling up" in subsequent decades.

Cybernetics and Big Ecology

Big Ecology is the term mathematical ecologist Daniel Coleman uses for his 2010 book describing the institutional frameworks that supported global "Big Science" projects in ecology. It includes international programs such as the International Biological Programme (IBP) of the 1960s and 1970s, sponsored by Man and the Biosphere (MAB), and increasingly from the 1980s it included the national programs such as Long-Term Ecological Research (LTER), sponsored by the US National Science Foundation (NSF).[65] Coleman writes with the perspective of a mathematically literate insider, an ecologist who was part of the teams working in the modeling era when ecology piggybacked on the computer revolution in the 1980s and scaled up to include the whole planet.

"Ecosystem science" was the most important framework for setting up large-scale and global research programs, and NSF funding followed this principle.[66] In a series of key papers across the 1940s and into the early 1950s, ecologists such as Raymond Lindeman, G. Evelyn Hutchinson, and Eugene P. Odum had developed the ecosystem concept into an abstract mathematical model of the energetics of *spaces* rather than focusing on the succession or physiology of species. The study of lakes (limnology) was particularly important for developing these ideas; because it was a more obviously bounded system it was easier to combine the concept of ecosystem with field results.[67] The most widely influential work on the energetics of the food chain came from Odum, whose work was initially funded by the Atomic Energy Commission through its Oak Ridge National Laboratory. Odum directly shaped the NSF's thinking. His *Fundamentals of Ecology,* continuously in print since the 1950s, with the latest edition in 2005, was an internationally important tool in the training of professional ecologists and in framing ecological "expertise."[68] Odum offered "a holistic ecosystem-oriented approach" to problems that was "a marked departure from earlier textbooks."[69] Because it was a book for students, it reached generations of ecologists at a time when they were most open to its ideas.

A crucial inspiration in Big Ecology was the mathematical biologist Alfred J. Lotka (see chapter 3), who was most famous among ecologists for his equations of the relations between *predator* and *prey,* highly influential as the discipline of animal ecology developed.[70] Lotka's 1925 book, *Elements of Physical Biology* (reprinted in 1956 as *Elements of Mathematical Biology*), introduced energetics and energy transformation to a wider audience and became crucial in ecological approaches that were amenable to large-scale computer modeling. His mathematical work enabled the multidisciplinary turn that global environmental sciences took in the latter part of the twentieth century. Thus, physics and mathematics—as well as the digital revolution— shaped the global projects of Big Ecology as much as ideas from the life sciences. This history also explains the alacrity with which environmental systems thinking was taken up to conceptualize the scalable behavior of the whole Earth. In the postwar era, the beginnings of Big Science in the biological sciences relied heavily on the NSF to bankroll them.[71]

The IBP marked the emergence of widespread mathematical modeling in ecology. Officially launched in 1964, it rescaled Tansley's 1935 ecosystem concept, taking it beyond local and regional applications to planetary scales from the 1960s. Frank Golley, another scientist who wrote a historical account of this era, was a director of research at NSF. The great success of IBP, in Golley's view, was not its capacity to meet the goals that were set by its original organizers ("the Biological Basis of Productivity and Human Welfare") but, rather, that in the long term it led to the institutional structures that supported permanent ecosystem studies.[72] By the 1980s, the global aspirations of IBP had narrowed to national lobbying for funds. The truly global aspirations of IBP became reduced to an international competition between the science-funding bodies of nations to hold "a place at the table of ongoing research," as Daniel Coleman described it. The outcome of renationalizing the biodiversity enterprise in the 1980s resulted in renewed efforts in the United States and its nationally strategic neighbors (particularly South and Central America), rather than a global approach. Despite the global aspirations of the new age of the environment, the management of practical research reverted to individual nations and smaller jurisdictions (see chapters 2 and 3).

Major conservation biology journals also still reflected national biases, not least because of research funding and teaching arrangements. A survey of the journals *Biological Conservation* (established 1968), *Conservation Biology* (established 1987), and *Biodiversity and Conservation* (established 1992) revealed that biodiversity research tends to be undertaken in the country of the author. It also revealed that most authors came from First World countries, and most often biological survey focus is on national parks and protected areas in those First World countries.[73] Thus, most of the work to protect species and ecological communities is being done in the places where biodiversity in a planetary sense is *least* threatened, a heritage too of the local practices of nature protection. Such literature surveys were alarming for a planet where typically global threats are to biota in developing world economies, in places not protected by biodiversity legislation. Thus, the history of ecological thinking is also a political history of how a concept enabled big-team expertise and facilitated funding rather than an intellectual history of discovery.

Conservation Biology as the Science of Crisis

In 1986, North American biologists Thomas Lovejoy, Michael Soulé, and Edward O. Wilson tackled what they saw as a "crisis" in conservation by broadcasting a new and media-friendly concept, "biodiversity."[74] This idea sought to put numbers on the dramatic loss of species. Rather than simply tracing losses of individual species as many reports had done, biodiversity enabled a global index for extinctions as general, even planetary processes. Their new metric could be turned into targets for new global nongovernmental organizations (NGOs) and private conservation initiatives, building on the techniques and systems thinking that had shaped global change thinking since the 1940s. It was an attempt by conservation science to deliver a general index by which the trajectory of ecologies could be measured holistically, something the various projects of Big Ecology had hitherto lacked. It echoed the proven influential tools for predicting resource availability, demography, and, increasingly, climate (see chapters 3 and 5).

The concept of biological diversity had a much longer history, but biodiversity in its short form was designed to enable nature to be "counted" so that its management could be supported by private conservation initiatives. One of the largest, Conservation International (CI), was established in 1987. Over the years, CI has issued tables of decline that are used to contextualize national and local conservation initiatives.[75] The world's top seventeen "megadiverse" nations include only two with a developed economy (United States and Australia). Comparisons can be made in terms of numbers of threatened species on the planet, using the International Union for Conservation of Nature (IUCN) Red List of Threatened Species and numbers of extinctions in different categories: "marine animals," "small mammals," "flightless birds," and so forth. Such numbers provide a good rationale for fund-raising to support ecological experts in the task of managing biodiversity conservation. This approach to applied ecology sought to manage biota, particularly species close to extinction, as a "crisis" caused by the actions of humans. Soulé described this work as "triage," by analogy with working in an emergency ward in a hospital.[76] His language unconsciously echoed Vogt's conclusion of 1948 that "the world is sick." Perhaps in a similar fashion to medical triage, the remedy is more focused on alleviating the symptoms rather than the underlying cause in the dynamics of society and human behavior.[77]

While the dramatic analogy with a medical crisis drew private sponsorship for conservation biology, other ecologists focused on numbers that translated nature into the economic systems of national and international politics in a bid to get public funding for what is essentially a public good. From the 1970s, increasing use of economic metaphors such as "environmental services" and "ecosystem services" suggested that accounting systems should no longer treat clean water and fresh air as "free" and that organizations polluting public goods should pay for the right to pollute.[78] Thus, by the end of the millennium, the sciences of ecology and conservation biology increasingly trusted numbers: for modeling, for translating management work into crisis frameworks, and for measuring the worth of nature to world economies. This work itself was made easier by the development of concepts and modeling that we have traced since the 1920s, when aspects of the complex web of life could be treated as "indicators" and fitted against the demands of disciplines with similar but separate trajectories (as discussed in chapter 3). Ecology was becoming increasingly overarching, important to policy making, and situated in spaces between disciplines and beyond ivory towers. It was no longer just a minor player in a botany or zoology department.

The biodiversity revolution offered a role for ecological expertise in big debates emerging about the "sixth mass extinction," a deep time perspective on the functioning of the Earth system allowed by increasingly detailed paleological data.[79] But the "biodiversity moment" of 1986 and the role of conservation biology as deliberately framed science of crisis was eclipsed almost immediately by the atmospheric chemists and the Greenhouse Summer of 1988, where the emergence of "unpleasant surprises in the Greenhouse" stimulated the foundation of a new body, the Intergovernmental Panel on Climate Change (IPCC).[80] No sooner had ecologists developed biodiversity as a way to measure and offer trusted numbers to manage nature in ways that ordinary people could understand than climate science experts suggested that the "reserves" put aside to protect nature "forever" were going to be ecologically changed by new climates.

Biodiversity loss has been identified as the most severe effect of all the global change events of the twenty-first century, using some four planets' worth of resources.[81] Ecologists can no longer work to address these challenges without collaboration with geophysical sciences, economics, and others in initiatives such as the Millennium Assessment and Rosetta Stone.[82] The challenge of systems ecology around 1950 was that it "called for a level of knowledge of physics, chemistry, geology, meteorology, and hydrology [and] required new skills in methods of instrumentation, technique and computation."[83] The practical effect was that no one person could manage this, and research shifted in the direction of Big Ecology, collation of data, and new "meta-specializations," especially in dealing with applied ecology at a supra-local level. The continuing challenge of environmental destruction and change left ecology—a leader in impulses toward disciplinary integration in the first half of the twentieth century—more of a junior partner in the meta-specializations by the century's end. Ecology has increasingly become a word to add on to other fields of endeavor—restoration ecology, political ecology, ecological economics, ecological humanities—rather than the expert science of nature's economy.

This is testimony, perhaps ironic, to the crucial role ecological thinking played in the emergence of the environmental age and the widespread appeal of ecological ideas, even if some of them, such as the old "balance of nature," have long been rejected by ecologists. It has forced environmental science to look beyond climate and altitudes as the drivers of vegetation (as biogeography had established in the early nineteenth century). Ecological systems depended on soil chemistry and mycorrhizal organisms, on the seasonal availability of water and light (not just the average amounts of these things), and on other drivers such as fire and the properties of microclimates. As the twentieth century wore on, ecologists increasingly recognized that human influence (history) was also a major part of understanding ecosystem structures. Such factors could be as diverse as agriculture and colonization, war, urbanization, use of pesticides and fertilizers, creation of protected areas, infrastructure as barriers and new routes for species, and many more. These factors applied differently in different places and were frequently not amenable to global or even regional averaging. The problem of the world beyond the "stable ecosystem" is that "chaos" (and chaos theory) has neither political appeal nor policy usefulness.

Analysis focused on conserving localized populations of particular species or iconic, endangered, and scientifically significant species had allowed science to be "policy relevant yet policy neutral." This catchphrase was later promoted by Robert Watson, chair of IPCC in the 1990s, as ecology crept into government policy in the postwar years. The more complex and dynamic the ecosystem, however, the more difficult it is to show that the advice is "neutral." The expertise needed to interpret graphs and data on many scales in relation to one another is hard to make transparent. As the global change story has taken on more and more dimensions, the details of local ecologies have become harder to translate. Scientific ecologists have found their expertise increasingly in conversation with the wider interdisciplinary environmental sciences and a wider range of relevant governing institutions (see chapter 6). They now work at juggling scales and policy directions while trying to theorize how revolutions in global climate science differentially affect local ecologies.

Climate Enters the Environment

Climate's Unpleasant Surprise

"We play Russian roulette with climate, hoping that the future will hold no unpleasant surprises." So warned Wallace S. Broecker in a commentary in the journal *Nature* in 1987.[1] Broecker (b. 1931), professor of earth and environmental sciences at Columbia University, New York, and a prominent oceanographer, had realized that paleoclimatological records showed that climate did not change gradually, as often previously thought. And it was already changing. He feared there might be bigger upheavals ahead—unexpected bullets hidden in the chamber of the gun. As he wrote, climate change was becoming a central issue for the planet; soon, perhaps *the* issue.

Broecker's warning was part of a deeper and more comprehensive shift of understanding going on across the emerging sciences of the Earth system. Broecker's work focused on radioactive carbon and carbon dioxide absorption by the oceans. Remarkable results were emerging in particular from ice core drilling—"the evidence that turned our heads." Ice cores extracted from the Greenland ice sheet provided a record of millennia of interactions between humans and the environment, revealing that, historically, climate did not change gradually. Scientists were just beginning "to see how devilish are the links between components of the climatic system."[2] Climate apparently changed in rapid spurts, the air temperature sometimes leaping up 6 or 8 degrees Celsius in just a few years in the colder regions, accompanied by major shifts in ocean currents. Such changes were visible to ice core specialists in the form of bubbles of air trapped in the ice, bubbles that contained shifting levels of atmospheric carbon dioxide.

Examining nature in narrow, separate categories, using different technologies, was no longer enough by the 1970s. We have already seen that new scales of analysis in ecology stretched the boundaries of the discipline itself. Similarly, the geological, atmospheric, biological, ecological, oceanographic, and cryospheric (for the sphere of ice and snow) sciences were all questioning the idea of slow and incremental (linear) change in nature. Together their insights showed that predicting the future climate was complex, dynamic, and nonlinear. Ecologists started talking about "disturbance" rather than "balance" as the fundamental property of the regime (see chapter 4). In evolutionary theory "punctuated equilibria," periods of slower change interspersed with periods of revolutionary upheaval, became a core concept. *Disaster* rather than *development*, *revolution* rather than *evolution* became the new ideas framing thought. When in 1999 French and Russian scientists presented a 420,000 year ice core from the Vostok station in Antarctica its data corroborated the Greenland story, but on a much longer timescale. Drastic shifts with immense consequences for all life on Earth were what we should expect when the level of greenhouse gases changes.[3]

Broecker's "surprise thinking" about climate would soon become pervasive. As the possibility of rapid climate change became apparent, there followed an enormous interest in concepts such as disaster or crisis and the corresponding terms that suggested how we might break out of the too-gradual change in economy and society: *transitions* and *transformations*. The Hollywood film industry, alert to the commercial dimensions of the ecology of fear, issued the spectacular film *The Day after Tomorrow* (2004). In hyperdramatized form, it presented the prospect of a catastrophic ice storm that paralyzes North America, with New York City turned into a glacier where only a few skyscrapers appeared above the ice line, like ancient nunataks of a misconceived modernity.

Of course one cannot attribute this whole way of thinking to one, albeit famous, article, or indeed one single scientist. Yet Broecker's surprise article is indicative of the cutting-edge science of his time. It reached beyond his immediate peers, enabling a major shift in both technical and popular thinking, affecting those only dimly or not at all aware of articles in *Nature*.

This mind shift from the expectation of slow, steady (linear) change is still ongoing. The world has not yet fully absorbed the idea of abrupt climate change. To acknowledge the possibility of such change would imply changes in many institutions, including insurance, security, and ideas of justice and the global distribution of risk. Such changes are so profound that denial, or at least downplaying the potential ramifications, is an understandable response. Sophisticated questionings of the relevance of mitigating actions or treating the possibility as an "unproven hypothesis" shield institutions deeply rooted in cultures, religions, psychologies of everyday consumption, and capitalist economic systems. Might coping with climate "change everything"?[4]

Broecker's article begins ominously: "The inhabitants of planet Earth are quietly conducting a gigantic environmental experiment." The actions of humanity could affect the entire planet. Such actions were unique; never before had this occurred, hence use of the word *experiment*. We have previously seen people suggesting the whole world had become a "laboratory" (see chapter 2). But this was a process beyond our control. Humans were no longer dispassionate observers of planetary processes but unhinged gamblers. Even more important for our story, the experiment was "environmental." To Broecker, the link between climate change and the environment was obvious; his use of it went without further reflection. In fact, one of the things he wished to achieve in his commentary was to draw attention to what he argued was a mishandling of "environmental research." In Broecker's view, it was overly influenced by government agencies and too little by universities that were free of political pressure, less prone to seek quick fixes, and more committed to basic research.

By the last two decades of the twentieth century, climate had become an environmental issue and climate itself was entangled in "the environment." Environmental research encompassed climate research. Today, as researchers on the environment, we often find that people assume this work primarily relates to climate. Yet this connection was far from obvious before the 1980s. Climate change science had a different trajectory since the nineteenth century and stayed outside the environment for a surprisingly long time in the twentieth. Climate and environment were nurtured in different scientific communities and politicized in very different ways. Both were entangled in issues of security and military concerns, but in ways that kept them apart. In the twenty-first century, by contrast, climate has become so much a part of the environment and predictions for its future that we run the risk of "reducing the future to climate."[5]

Airs, Waters, Places

Climate was until very recently considered a local phenomenon. Terms like *microclimate* (i.e., the local part of a global system) only emerged once climate itself was scaled a global phenomenon—that is, very recently. *Klima* had been used in antiquity to signify the totality of physical conditions that nature provided in particular places. Ancient Greek astronomers and geographers divided the world into zones or parallels, and *klima* comes from the verb *klinein*, "to lean, recline." The Earth "sloped," the Greeks thought, from the equator to the North Pole, and they were aware that this sloping caused different weather conditions in the respective regions. Hence, each of the regions in the world had a *klima*, corresponding to its "inclination." This explains the root meaning of climate as something fixed and given to a place rather than something that might change. This geographical determinism of climate was used in Greek thought to speculate on the equally fixed and place-based relationship between climate and character. The texts *On Airs, Waters, and Places*, from around 400 BC, attributed to Hippocrates, a Greek philosopher and physician, considered that northern climates made people slow and dull, while inhabitants of humid climates got coarse skin and retarded minds, and heat made it hard for Africans to think. By contrast, Hippocrates argued, Greece's mixed and temperate climate conditions were ideal for work, for sound judgment, and for the maturation of wisdom. His was the first of many philosophies of climatic determinism that were still proposed in ever-new versions well into the twentieth century (and usually lauding the climate that the author liked best!).

Climate was about *the register of place* and what it implied for the future of cultures and peoples in those places. Climate was a local concept; different places and regions had different climates, which was also why people and societies were different. In fact, this meaning was very close to the concept of "environment," as the circumscribing conditions that affected "man" and his actions, as it emerged in English during the nineteenth century. "Environmental protection," before 1948, consequently meant the protection not of the broader environment "out there" (the nature that humans affect) but instead a *protection from* environmental forces working on humans. "Environment" in this tradition, common in physiology and psychology, meant and still means precisely the local conditions that influence humans, often with an emphasis on the pressure and stress they cause (see chapter 2). Environmental protection was about clothes, food, and the know-how of well-being and survival under surrounding conditions that could maintain stability in what a French nineteenth-century physiologist called *milieu intérieur* of the human body. Environmental physiology and psychology were equally applicable to microenvironments, such as the interior of a building or anything larger. This was the thrust of studies commissioned by the US military and undertaken by, among others, the Harvard Fatigue Laboratory in the 1930s and 1940s on "temperate areas," arctic and subarctic areas, mountain areas, desert areas, jungle areas, or "moist tropical areas." These were considered the "environments" of soldiers who would perform differently under different climatic conditions.[6]

In these studies focused on people and their geographical surroundings, the dominant understanding of environment still derived from the Greek idea of climate, although they increasingly used the word popularized by Herbert Spencer and a range of ecologists and social scientists. Any region has a climate that influences cultures and characters. Environmental physiology and psychology are radical versions of this idea, scaled to individuals and their immediate surroundings. Climate was an exclusively local phenomenon, a given that humans could not change. They had to protect themselves against it to stay functional, for example, in war and to perform in work, school, sports, and other pursuits. As we have set out, the emerging understanding of environment, 1948-style, was completely different. It was about humans changing and threatening the environment on all scales, from the microlevel to the planetary. Climate, as yet, had no place in this story.

The local and regional scale of climate was taken for granted into the second half of the twentieth century. Climate only started to denote a *global* condition recently, and with a broader understanding of its spatial span came an increasing understanding of its deep time history, particularly through reading ice cores and the study of glaciers, although some climatic changes of the Holocene (the geological epoch spanning the past 11,700 years) had been known for a long time from other sources such as glacial retreat, paleo-botanic records, tree rings, and sediments.

From Local to Planetary Climate

Today it is hard to conceive of the environment without also thinking about the global climate. Yet, only in the 1960s did climate come to be considered an environmental problem, and it was found rarely in research documents even then. One example was in an appendix to the report on environmental pollution delivered in 1965 to President Lyndon Johnson, *Restoring the Quality of Our Environment*. A product of a study conducted by the President's Science Advisory Committee, it analyzed the status of the environment from many angles, mainly concerned with chemicals and toxins but also waste, "metropolitan problems," and climate. As the title of the report indicated, the "quality" of the environment was already perceived to be so bad that it needed "restoring." The issue was presented as one of primary importance. "Our country's continued strength and welfare depend on the quantity and quality of our resources and the quality of the environment in which our people live."[7] Wallace Broecker served on the study's subpanel for "Atmospheric Carbon Dioxide" along with, among others, Charles David Keeling and Roger Revelle, who were both to play key roles in the modern discovery of anthropogenic climate change.

The understanding of global "climate" and its inclusion among other environmental problems was not a process that occurred simply because of enhanced scientific understanding. Nor was climate reshaped and integrated into other considerations only by science. To begin with, climate science as a distinct field was established surprisingly late, largely during the latter decades of the twentieth century. Almost all scientific journals with the word *climate* or even *climatology* in their title date from no earlier than the 1980s, in some cases the result of renaming earlier journals. *Climatic Change* (established 1978) was one of the first to have that key word *change* in its title, as a new orthodoxy was established. Scientific research and reflection on climate and its role in human societies is of course much older, and a notable example that did consider change was founded by German geologist Eduard Brückner in 1906 under the auspices of the International Glaciological Commission: *Zeitschrift für Gletscherkunde: Für Eiszeitforschung und Geschichte des Klimas* (Journal of Glaciology for Research into the Ice Ages and History of Climate). Brückner was among the few in his time who regarded broad-scale changes in climate as possible and possibly even having human origins.[8] The study of the advance and recession of ice sheets and glaciers provided, of course, prima facie evidence of climate change. But this journal and the entire Glaciological Commission fell into disarray during the interwar years and never recovered.[9]

Nevertheless, there were other significant intimations of thinking about the planet as an interrelated whole. German geographer Alfred Wegener in 1912 formulated a theory of *continental drift*, whereby the continents themselves moved apart, isolating biota from their shared deep past. But the mechanism for how continents drifted was lacking, and his idea was not widely established until the 1960s when the theory of plate tectonics provided the explanation through sea-floor spreading, where new oceanic crust forms through volcanic activity, then moves away from the mid-ocean ridges.[10] Oceanographers also brought together insights concerning deep-sea currents that connected oceans and continents, while meteorologists and other geophysicists studied airstreams flowing over long distances, providing a pattern of geophysical motion.[11] Such a planetary understanding derived from new technologies and infrastructures of monitoring, travel, and communication, including air travel.[12]

In this book, we have also encountered earlier intimations of change on a planetary scale from thinkers such as Georges-Louis Leclerc, the Comte de Buffon, in the eighteenth century and Eugène Huzar in the nineteenth. Buffon and Huzar entertained the notion of humans "improving" climate, an idea evident since early modern times. Their understanding was, however, largely local or regional. Buffon compared the climate of densely populated Paris with that of Québec on about the same latitude; the former had improved thanks to the large-scale presence of humans. Deforestation in the Americas, for example, was thought to have caused changes in temperature and later would be blamed for localized desiccation. George Perkins Marsh's *Man and Nature* of 1864 was another grand, empirically detailed account of how Europe and North Africa over centuries and millennia had been transformed by human culture and ingenuity. Although it devoted little attention to climate, the issue of localized desiccation was widespread in the nineteenth century.[13]

Around 1900 the idea that it was possible for humanity to influence *global* climate began to gain traction. It was, after all, a very radical idea. In 1896 a now-famous calculation of how much global temperatures would rise if humans burned the then-known coal and oil reserves was made by physical chemist Svante Arrhenius at Stockholm University. He regarded the planetary "greenhouse" as a thought experiment. He never expected it to happen. Other influences came from cosmology and the contemplation of the development of life over the vast timescales revealed by nineteenth-century geology. Russian scientist Vladimir Vernadsky, who lectured on geochemistry at the Sorbonne in Paris during 1922–23 as part of the international intellectual cooperation after the Great War and the Russian Revolution, contributed significantly to the idea of anthropogenic change. His lectures were published as *La géochimie*.[14] In a subsequent round of lectures, he coined the concept *biosphere* in his book *Biosfera*.[15] The biosphere described the slim zone of organic life and atmosphere wrapped around the Earth's crust as an integrated whole. He was already convinced that people were eroding their own "survival" by wasting "parts of the biosphere which provide the things that *Homo sapiens* as a mammal and as an educable social organism needs or thinks he needs." Vernadsky had a global, or rather planetary, approach that was required to conceive of global climate as something vulnerable. As population expanded, wasting accelerated, a point that would be made in the context of postwar Malthusianism by Yale ecologist George Evelyn Hutchinson, who translated some of Vernadsky's ideas and started introducing them to a wider audience.[16] Hutchinson and Vernadsky both made the argument that human activities could affect the biosphere on a grand scale and that they had already done so through mass extinctions and the emission of greenhouse gases. They were among the heralds of modern environmental understanding, but even more, they were adamant that all life, as one big entity, shaped geological evolution.

Vernadsky was a mineralogist and biogeochemist rather than a climate scientist. His ideas of human impact on climate were more philosophical than derived from his empirical work. His idea of an atmospheric *greenhouse effect* was inspired, in fact, by Arrhenius, whose theory he integrated into his own notion of the biosphere. Arrhenius in turn was preoccupied not with climate change in the present or near future but with the causes and consequences of the Ice Ages. Most scientists thought this problem was resolved by the 1920s as a result of the discovery of the Milanković cycles (see page 107), wobbles in the trajectory of the Earth around the sun that affected the flow of incoming solar radiation and hence global temperatures. It was merely as an aside that Arrhenius also postulated that the combustion of fossil fuels would raise carbon dioxide levels and cause global warming, a positive prospect for a northerly nation like Sweden.[17] Despite Hutchinson's early efforts, he had to reiterate his dependence on and admiration for Vernadsky in 1970 in his introduction to a special issue of *Scientific American*, "The Biosphere," published in the context of the discourse of Earth Day and the rising American environmental movement.[18]

Early ideas of anthropogenic climate change were also challenged by those who opposed climatic or environmental determinism, which claimed that human societies or races were the outcome of local climates. The argument was simply that if you believed in free will, you must allow for humans to have a choice, to change their thoughts and their actions, and hence no prediction of human behavior or future climate was possible. In 1929 Arthur Eddington argued that if human-made climate change existed, no physicist could predict the weather one year ahead. How could one advance any further? Equally, he asserted that while the human mind guided human action and could alter the chain of events on Earth, physical determinism is impossible. "Therefore," he wrote, "we must penetrate into the recesses of the human mind. A local strike, a great war, may directly change the conditions of the atmosphere; a lighted match idly thrown away may cause deforestation which will change the rainfall and climate. There can be no fully deterministic control of inorganic phenomena unless the determinism governs mind itself."[19] If humans had a choice, then they also had responsibility. Eddington's position was considered marginal at that time. He focused mostly on ameliorating local climates, not planetary-scale climate change, and opposed the determinist tendencies that were growing strong in the interwar period, both in racist politics and in certain branches of geography, anthropology, and medicine. Such theories clashed with ideas of human freedom.

Converging Discourses

The story so far traces the discourses of climate and the environment but not how they converged. Theoretically, geophysical scientists were aware of the possibility of people's actions affecting climate since Arrhenius wrote his 1896 paper. With the notable exception of Vernadsky, however, they avoided the idea of anthropogenic climate forcing.[20] Some concerns about climate change in the form of "desiccation" were revived after the 1930s dust bowls in the United States and Australia, but any anthropogenic element of these theories of climate variation was vague.[21]

The idea of measuring the anthropogenic forcing of global climate as an ongoing, real-time phenomenon came from an unexpected source. Guy Stewart Callendar, a British steam engineer and expert on fog formation, suggested in a paper read at the British Meteorological Society and published in 1938 that human burning of fossil fuels had already led to an increase of the atmospheric content of carbon dioxide and a discernible, indeed quantifiable, rise in global mean temperatures.[22] He repeated his claim in a string of papers during the following decade. The scientific establishment of the time distrusted his ideas and perhaps just as much his calculations, which were made in his spare time as an amateur climatologist, lacking the prestige and infrastructures of a major institution.[23] In fact, by mid-century, when the new formulation of the environment started to gain traction, the idea of anthropogenic climate change was stone dead, discarded in meteorology textbooks as an old hypothesis proven wrong. It was not mentioned in William Vogt's 1948 book nor in the following years despite a rich outpouring of environmental jeremiads and scholarly books and papers. The story of the gradual acknowledgment and finally widespread endorsement of Callendar's work—a story that unfolded over several decades—is therefore also an account of how the gap narrowed between the two discourses of environment and climate.

In a few corners, however, interest in levels of atmospheric carbon dioxide and the possibility of its impact on climate lingered. It floated around in small scientific circles that were aware of Callendar's work. In Scandinavia, Stockholm-based meteorologist Carl-Gustaf Rossby, with a long career behind him at the US Weather Bureau and at the Massachusetts Institute of Technology and the University of Chicago, took up the challenge.[24] In the early 1950s, he was in conversation with Finnish chemist Kurt Buch, who had developed an interest in apparently increased atmospheric carbon dioxide based on data that Buch had collected for a long time in the north Atlantic. The Scandinavian scientists, under the leadership of Rossby, formed a network of stations to monitor atmospheric carbon dioxide from 1953.[25]

From around this date the wider media became interested in the possibility of human-induced climate change through the combustion of fossil fuels.[26] Soon after, anthropogenic climate change became part of the research program for the International Geophysical Year (IGY) of 1957–58. In 1956, Rossby was quoted in a *Time* magazine interview as saying, in precisely the words that Broecker echoed three decades later, "mankind is now performing a unique experiment of impressive planetary dimensions by consuming during a few hundred years all the fossil fuel deposited during millions of years. . . . There is no doubt that an increase of carbon-dioxide content in the atmosphere would lead to an . . . increase of the mean temperature of the atmosphere."[27] The same year, meteorologist Gilbert Plass published a seminal paper on the carbon dioxide theory, endorsing it as very plausible.[28] This was a turning point, and in the following years a stream of papers discussed the possibility of an actual global warming because of the anthropogenic rise of carbon dioxide, a rise that no one denied and everyone predicted would continue. Some questioned the precise causal relationship, based on the absorptive capacity of the oceans, while others thought of it as more or less a self-evident fact. Plass also forecast the carbon dioxide increase in the atmosphere up until the year 2000, quite accurately as it turned out.[29]

Yet it was not an "environmental" issue. The communities of scientists that started developing a language of environment, based in ecology, conservation, geography, resource economics, and other fields, were quite distinct from the community of geophysical scientists. Even the single community united by the IGY was divided into geologists, meteorologists, atmospheric chemists and physicists, oceanographers, physical geographers, glaciologists, and space scientists. Each of these groups had preferred lines of reasoning and different theories about changes in climate and their cause. It was also a community with very little interest in conservation and ecology, and they had no expertise whatsoever in the workings of humans and societies. Astronomers remained apart from debates about human climate-forcing, although their interest had been engaged by Croatian scientist Milutin Milanković's theory of multimillennial climate change relating to solar cycles, which attracted attention in the 1920s.[30] The physicist Fred Hoyle, creative and controversial and engaged in population debates, supported theories of climate change focused on a "change of the solar constant," an idea he had presented in 1939. Hoyle's approach was simply to state that "we shall never know" what might have been the impact of the sun from its galactic passage through clouds of interstellar gas over a matter of a "million years," unimaginable on the human timescale, but he admitted that change over periods of a thousand years was detectable.[31]

Other groups were more engaged. The US military and their supporting scientific and funding agencies already used "environment" as a guiding and operational concept in the 1950s as they gathered intelligence on the conditions for Cold War geographies, from the sea floor to the stratosphere.[32] They were also interested in possible climate change and what might be learned from it to benefit the planning of warfare, especially in colder climates. But as already noted, they were not environmentally minded. They had little interest in the environment as something that was threatened by human activity. Quite the contrary; however, they had money.

Mobilizing Climate and Environmental Warfare

The Second World War had mobilized the various geophysical sciences and collaborations between them. This continued and was reinforced during the Cold War. The concept of "environmental warfare" referred to the use of environmental factors as an indirect weapon, for example, the use of vermin, insects, or disease.[33] Weather or even climate manipulation belonged in this potential arsenal, for instance, through the hydrogen bomb. John von Neumann, the Hungarian-American mathematician, made his name as a hard-nosed Cold Warrior by discussing seriously the strategic possibilities of making a "new ice age," melting the Greenland ice sheet and reversing the Gulf Stream. To describe the extraordinary potential of "climate control," he invoked, with no concern at all, the word *environment*. "What power over our environment, over all nature, is implied!"[34]

That the Soviet Union was planning to block the Bering Strait to manipulate climate was considered a realistic assumption in North

Atlantic Treaty Organization (NATO) war planning circles of the 1950s and 1960s.[35] NATO also conducted massive environmental research, including work completed through military funding agencies in the United States and its allies. Computer-based environmental predictions were engineered into the Cold War scenario planning that became the hallmark of the RAND Corporation (a government-funded research body and exemplar of the "military-industrial complex" as coined by President Dwight D. Eisenhower in his farewell address in 1961) in Santa Monica, California, with climate as a main factor. RAND scientists suggested blackening polar ice sheets to increase the world's temperature by preventing energy from radiating out. In what for a moment seemed like a science fiction playground of climate change speculation, scientists also proposed dropping nuclear bombs on ice sheets to make them slide into the ocean, creating tsunamis but over the long term inducing widespread Arctic chill as the reflective areas of the Earth's surface might be extended.[36]

The geophysical sciences, regardless of whether they had a military or civilian orientation (the boundary was not always easy to discern), generally were slow in adopting an "environmentalist" perspective. Yet as they accumulated evidence, they would also be essential in providing evidence of environmental change and in making the environment a factor of utmost importance.[37] There was even some fear that producing scenarios and predictions of a postnuclear or post–hydrogen bomb world would favor the emerging environmental movement, which had started generating its own doomsday scenarios. Still, the perspective of weather and climate engineering was far distant from the tradition of care for the planet that William Vogt, Fairfield Osborn, and others had launched. These early advocates had regarded the environment as the increasingly weaker or "fragile" partner of a lasting relationship that both parties would lose if humanity, growing in numbers, treated her badly.

In contrast, for the military and their support structures in government agencies, universities, civil research institutes, and think tanks like RAND, the environment was instrumental, a possible source of solutions to a military problem. It was merely a weapon or tool, not worthy of moral consideration. In certain scenarios it was even something that could be fully controlled, which required only more active interventions and terraforming, not less. Or, as von Neumann put it, "the environment in which technological progress must occur has become both undersized and underorganized."[38] Caring for nature was irrelevant to what was required for warfare or even "progress." Such instrumentalism survived in later discussions of geoengineering, a civil form of "environmental warfare." Typically, there was another end of the spectrum, where concerned Cassandras whispered their warnings from the humanities. Such technological overreach was all the fault of war, argued the historian Lewis Mumford, and of states who listened to the wrong advice. Others blamed wars themselves on overpopulation and competition for resources (see chapters 1 and 3).[39] Lynn White Jr., medieval historian and theologian, perceived an "ecologic crisis" caused by technology with its roots in Judeo-Christian beliefs in the domination of nature.[40]

Cold War military interest in climate extended over many geographies, from deserts and the tropics to the polar regions. The Arctic was a potential theater of war where the "superpowers" were also geographically closest to one another. Both the Soviet Union and the United States built a formidable capacity in sophisticated snow and ice research.[41] The American "Project Iceworm" built an entire town, complete with a nuclear reactor for energy supply, in the Greenland ice cap.[42] Arctic submarines were developed with a capacity to rise through thick ice sheets.[43] Ice started to play a role in the understanding of the human predicament. Canadian scientist Graham Rowley began measurements of sea ice immediately after the war, complementing those that had been provided by Russian and Norwegian scientists since 1900 and accelerating in the interwar years.[44] To this must be added the enormous local knowledge of northern populations. Ice core drilling had started around 1950 and in the 1980s and 1990s the impressive Greenland and Vostok ice cores (Antarctica), extracted by Americans and Russians, respectively, added significantly to climate knowledge and, as we have seen, revolutionized its empirical status.[45]

Military technologies were not limited to the surface of the Earth. With the arrival from the early 1970s of satellites, a much-enhanced overview of the globe was enabled. This gave a novel perspective on the polar regions, too, and the presence of ice at different elevations. The concept "cryosphere" became increasingly used to describe this frozen world, seemingly as an addition to the list of concepts starting with the geosphere (used by Aristotle), atmosphere (since the seventeenth century), and hydrosphere (nineteenth century) but perhaps more particularly as an addition to the rapidly growing list of "geospheres" or "envelopes," according to principles proposed by the Swiss geologist Eduard Suess in 1875.[46] The term *geospheres*, in the plural, was coined in 1910 by the oceanographer John Murray, who had traveled around the world with the marine research vessel *Challenger* and acquired a vivid sense of the global. *Troposphere* and *stratosphere* were introduced by Leon Teisserenc de Bort in 1902 and *asthenosphere* by Joseph Barrell in 1914. It was in this tradition that we can also place Vernadsky's *biosphere* of 1926, mentioned earlier.

Cryosphere fits this list as well, although just like biosphere it had its real breakthrough only recently. The first use of the concept was by the Polish glaciologist Antoni Dobrowolski in 1923 in a massive volume entitled *Natural History of Ice* (in Polish).[47] Dobrowolski may have written parts of this volume several years earlier in Sweden, where he spent time during the First World War. However, the concept of a cryosphere as an analytically useful space was vehemently protested by glaciologists and geographers, most of whom were equally skeptical of the idea of anthropogenic climate forcing. The concept was largely forgotten until it returned in the 1970s through the global view of satellites that showed widespread snow and ice (no less than a quarter of the Northern Hemisphere in January) and through the computer-based general circulation models that had started to model the reflective albedo (literally meaning "whiteness") effect of ice that corroborated the idea of anthropogenic climate change. Suddenly the extent of snow and ice became a key indicator of changes in the climate—perhaps even of all our fates. Melting ice became the symbol of global warming par excellence. The fractures in polar

upon nature and climate is a system of scientific procedures which the people have named Stalin's Plan for Reforming Nature."[49]

Thornthwaite was clearly skeptical of schemes of this bewildering magnitude, just as he disbelieved the necessity for George Perkins Marsh's call for prudence in relation to nature. We may be able to change *the face of the Earth*, but humanity cannot change climate, argued Thornthwaite in 1955, despite increasing signals that this very thing was possible.[50] In several other contributions to the conference, scientists advocated large-scale geoengineering projects to manipulate precipitation or the climate of towns.[51] But these were schemes akin to clearing forest for agriculture or damming rivers. There was no real sense of the global or of the planetary climate. Carl Sauer, one of the convenors, agreed with Thornthwaite, a former student, in arguing that humanity could not alter global climate. Carbon dioxide was mentioned a mere four times during the entire conference, and none of these instances referred to anthropogenic climate change. Had the meeting been held only a few years later it would probably have been on the agenda; it was precisely the middle years of the 1950s that marked the discrete and silent breakthrough of a wider understanding in relevant scientific communities of the modern orthodoxy on this issue.

In the following years, computer power revolutionized work on climate modeling. In the middle of the 1960s, the first computerized general circulation models, later just "GCMs," started to appear, and from the outset they projected a drastic increase of global atmospheric mean temperatures if burning of fossil fuels continued at then-present, rapidly growing levels. Groups of geophysical scientists became increasingly concerned and started to develop comprehensive research programs on the matter. At this point, the rise of the discourse on the environment, including popular anxieties, meant more for climate change scientists than the other way around. In 1970, when the Environmental Protection Agency was formed in the United States, "environment" had become the concept of the day and planning was under way for the UN conference in Stockholm in 1972 (see chapter 6). Although previously well resourced by the military, climate science could now ride this rising tide. One of the major research initiatives in anticipation of the Stockholm conference was the project "Man's Impact on the Global Environment," conducted at MIT in 1970. It included the study of greenhouse warming, which at least by this group of scientists was now considered a relevant framing of climate change. Part of this reasoning had to do with the kinds of environmental *consequences* that might follow from rising temperatures: "widespread droughts, changes of the ocean level, and so forth."[52]

These kinds of environmental problems were typical concerns but, added to the prospect of global change, the prospective *scale* of impacts became significantly larger. Only rarely would pollution or overgrazing or overfishing yield such comprehensive damage to anything from ecosystems to grasslands to oceans to human health. This issue of scaling may be useful to understand the hitherto somewhat puzzling mismatch between climate and environment. This worked in two ways. On the one hand, once climate included more than the local level, the concept "environment" also moved to the scale of planetary *dynamics*, not just a set of issues found all over the planet. On the other hand, the discourse of the environment helped frame climate change as an issue of general concern, not just of esoteric interest. The global nature of the atmosphere and the rapid circulation of gases in it demanded models that were also global in scope. The global scale was equally important for understanding oceans and their currents, critical to creating conditions for climatic shifts like the El Niño Southern Oscillation (ENSO). Indeed, the air/sea exchange of carbon dioxide was the focus of work at the Scripps Institution, San Diego, led by Charles Keeling that, during the 1960s, provided the first clear empirical evidence of rising carbon dioxide levels.[53] Other areas of environmental science where *land* was typically the basis for conceptualization, empirical work, and modeling remained within local, national, or regional scales.

Thus, environment and climate worked in a complicated tandem. On the one hand, they mutually reinforced each other in the process of gaining purchase as urgent scientific and policy issues. On the other hand, they demanded different types of expertise, involved quite distinct scientific communities, and established themselves in different institutions, not only in universities but also in government agencies and, perhaps most distinctly, in relation to the military. Only after the 1980s did climate research become more integrated with other strands of environmental research and vice versa. Climate science did take the environment into consideration, but only under certain circumstances. As we have seen, the security interest in the environment had little to do with human impacts and was far more about environmental conditions for warfare and for the performance of soldiers and equipment.

More important for the long-term direction of climatic science was the emerging propensity among some of its practitioners to align their findings with emerging environmental paradigms. The observations produced by Charles Keeling at the Mauna Loa Observatory in Hawaii, established during the IGY, illustrate this point. Keeling supplied the first annual set of data on carbon dioxide in the atmosphere at a location generally free of localized interfering factors and that could be considered representative of conditions in the wider atmosphere. From the inception of the data series there was a clear upward trend, and indeed this long-term record of rising carbon dioxide concentration, key evidence for theories of anthropogenic global warming, is now known as the "Keeling curve." By the time he was speaking to a symposium on the long-term implications of atmospheric pollution in 1969, Keeling could confidently present his Hawaiian data to show the continuous rise of carbon dioxide.[54] That increase in and of itself meant nothing to most people. "But CO_2 is just one index of man's rising activity today," he argued, pointing to its equivalence to data on population growth, consumption of fossil fuels, smog and air pollution, and the clearance of virgin lands.

Keeling's 1969 address oscillated between optimism and despair. He drew on the geochemist Harrison Brown's dystopian *The Challenge of Man's Future* of 1954 (see chapter 2), which traced the long history from a nature in balance to a nature unbalanced by human activities. He also explored economist Kenneth Boulding's reformist agenda in *The Meaning of the Twentieth Century* (1964), where that period is portrayed positively as a transition from a civilized to a "postcivilized" society. Keeling positioned his ideas about the environment somewhere in between. Keeling foresaw a world in year 2000 "in greater immediate danger." People living then will, in addition to "their other troubles," also "face the threat of climatic change brought about by an uncontrolled increase in atmospheric CO_2 from fossil fuels."[55] The essence of his presentation was the fact that fossil fuels and environment were inextricably linked. The fossil fuels were changing climate, and climate was changing the human environment. Beyond geophysics, warfare, and debates about determinism, climate change had, with the irrefutability of the Mauna Loa Observatory curve, placed itself among the other anthropogenic threats to nature. Climate, or in reality humans and their societies driving climate change, was a co-creator of the environment on a grand scale. This was the great unpleasant surprise to emerge from basic curiosity-driven research. But importantly, there was already an environmental story in which to fit this new narrative of alarm.

Global warming issues were included in the UN environmental framework by the early 1970s. Under strong encouragement from Swedish Prime Minister Olof Palme, Sweden's permanent representative to the United Nations, Sverker Åström, elaborated upon the nature and extent of the urgent threat posed by environmental degradation, drawing heavily upon Swedish scientific knowledge.[56] One of the important preparatory events for the Stockholm conference, the Study of Man's Impact on Climate (SMIC), was held on an island in the Stockholm archipelago in 1971. SMIC became an early milestone on the road to forming a scientific consensus on the incipient threat of climate change. It generated a report published by MIT, which co-hosted the event together with the Stockholm-based Academies of Science and Engineering Sciences.[57] Here national traditions of scientific expertise mattered and continued to matter. For instance, the International Meteorological Institute at Stockholm University, founded by Carl-Gustaf Rossby in 1948, led the climate change assessment for the landmark UN report of the World Commission on Environment and Development, *Our Common Future*.[58]

Much of the Swedish climate science-policy interface involved the long-time director of that institute, meteorologist Bert Bolin, who was an indispensable actor on the international stage who deftly operated between the realms of science and policy. Among his institution-building credentials, Bolin was the founding director of the Committee on Atmospheric Sciences (1964), the Global Atmospheric Research Program (1967), and most significantly in 1988 the Intergovernmental Panel on Climate Change (IPCC). He also had good relations with the Swedish government, serving as scientific advisor to the prime minister. Through the influence he held in the latter position, he secured funding from the Swedish state to facilitate the establishment of the International Geosphere-Biosphere Programme (IGBP), headquartered in Stockholm, to coordinate international research on global-scale and regional-scale interactions between Earth's biological, chemical, and physical processes and their interactions with human systems (see chapter 6).[59]

The 1970s saw a breakthrough for the new orthodoxy. Paper after paper endorsed the understanding that Callendar, Plass, Rossby, Revelle, and others had projected. Potentially catastrophic scenarios were mooted. In 1971 it was suggested—as part of the Stockholm SMIC meeting—that the entire Gulf Stream might change its course. Toward the end of the decade, the National Academy of Sciences received a report (1979) from the National Research Council, which had appointed an ad hoc working group chaired by Jule Charney. Charney was an eminent MIT physicist and expert in applied mathematics whose entire career had been devoted to numerical weather predictions and to climate change science, where he was an advanced and sought-after modeler; he spent several years with John von Neumann's computer project at the Institute for Advanced Study in Princeton in the early 1950s. His report, endorsed by the National Research Council and acknowledged by the National Academy of

Sciences, served as the ultimate evidence that the established parts of the scientific community had now accepted anthropogenic climate change as the new norm.[60]

As with the emergence of Big Science in ecology, the study of climate change became scaled up institutionally in projects such as the Global Atmospheric Research Program, GARP I (from 1967) and GARP II. These mega-efforts of data collection and processing built the context for the later IPCC, part of a growing accumulation of data that had jump-started on the global scale with the IGY in 1957–58. Although it seemed that oceans and forests absorbed much of the carbon dioxide in the atmosphere (which, we remember, was the focus of Keeling's research), there was enough remaining in the atmosphere to convince most atmospheric scientists that it caused warming. Callendar's predictions, once discarded as hopelessly amateurish and plain wrong, were increasingly regarded as mainstream.

In 1988 the United States was hit by high temperatures in what became known as "the Greenhouse Summer." Giant fires burned in the iconic Yellowstone Park, and NASA scientist James Hansen testified to Congress that "The greenhouse effect . . . has been detected and is changing our climate now." The then-recent Montreal Protocol to stop ozone depletion suggested a template for addressing climate change. The Toronto World Conference on the Changing Atmosphere was organized that same year and the assembled scientists urged governments to set targets for greenhouse gas reductions, just as there had been targets set for scaling down on aerosols.[61]

The "greening" of climate change also meant that it was no longer seen as a predominantly benign phenomenon. The expectation of increasing temperatures as a benefit, as "amelioration," dated back to antiquity. There had been a handful of exceptions to this, such as fear of desiccation, which might drive "Asian hordes" into Europe, and general fears of desertification, but the overall discourse about climate change was not alarmist. Arrhenius, his fellow Swede Nils Ekholm, and their northern colleagues, on the contrary, suggested that the burning of fossil fuels might be a way of slowing the arrival of new ice ages. The discourse in the 1940s and 1950s was "scientific" in the strict sense, an attempt to figure out whether there *was* climate change, with all available methods. Atmospheric physics and chemistry then had none of that predictive prestige they would gain when GCMs came of age in the 1970s. By the 1990s, the reports of the IPCC were also prognostications on the future fortunes of humanity.

The Greening of Climate Change

The climate issue came to a climax in the first decade of the twenty-first century. Climate had grown so big that other dimensions of environment became marginalized in the popular media. Yet the environment still embraces much more than climate change, and climate concerns themselves drive sectors like "energy transitions" and insurance risk predictions, not just "the environment." Generally, climate change now adds to the ongoing expansion of environmental policy making rather than being separate from it. Climate is, in this sense, the best evidence that environment is becoming an integrated part of what Hannah Arendt famously called "the human condition." Climate, at least since Hippocrates, was considered the most fundamental of human conditions; airs, waters, places—what could be more of the essence for human existence? Another word for these elements is, precisely, *environment*.

Climate was absorbed into the environment through a long and circuitous process, starting in earnest in the 1920s when both concepts began to emerge in their modern meaning and accelerating through to the early twenty-first century. This chapter has thus followed a pattern already established in previous chapters where certain notions of societal growth (for example, population, economy) enter the ambit of the environment. Climate, although it is not about human societies, shares many of the features of economy and population as it becomes a relevant dimension of the environment by changing it, in worrying directions, at a worrying pace. Knowledge of the direction and rate of climate change seems to be a prerequisite for understanding the environment in the present century.

With climate, however, this played out somewhat differently. The changes in climate are not very easy to perceive or measure. Even more opaque to the layperson are the possible causes of this change, once its direction and rate have been established. This could only be done through computer programs and sophisticated monitoring on both the local and the global scale,[62] skills that were acquired later (although not very much later) for climate than for population or even economic growth. Nonetheless, they belong to the same family of phenomena. There may even be a certain kind of mimicry going on, including the methods of the digital revolution. The overarching idea that rates of change, established numerically and with a systematic trend or direction, are necessary for the construction of the environment holds as true for climate as any other dimension.

Climate change also brought with it new strands of expertise for the environment. In common with older environmental expertise, it was largely scientific and relied heavily on quantification. But its use of models of global systems was greater, and it had a far more tenuous connection with field stations, especially compared with early environmental sciences. Already, field experts like Vogt, who drew heavily on his ornithological fieldwork in Latin America, had been replaced by systems experts like Eugene and Howard Odum. Climate change science reinforced this tendency. Still, the scientific culture of atmospheric science was different from the already established "environmental science," which did not facilitate their entry; leading climate scientists such as Rossby, Charney, Plass, Keeling, and Revelle had conspicuously meager exchanges with life, field, and biological scientists, and vice versa.

Global programs and institutions reflected the new expertise. By the 1970s some of the earliest and the biggest global research programs, such as GARP I and GARP II, were related to climate change science.[63] As "global change" science programs started to appear in the 1980s, climate was a distinct and integrated part, signifying that it was at home in the environment (see chapter 7).

Thus climate has become a very important part of the globalizing narrative of environment. Even more important, climate globalized the environment through the very nature of climate change, as an increase of carbon dioxide in the atmosphere exists everywhere and causes a universal, albeit geographically varied, rise of temperature. When it was mentioned in the 1950s, it was merely one of several aspects of environmental change in local places, not a global issue. Climate went from being an ultimately local phenomenon, which it had been since antiquity, to becoming ultimately global. As such it is also unique, as it brought humanity together to face a common problem rather than just a local problem with parallels elsewhere. So climate change became an arena where the global was an obvious condition of the discourse, spurring the environmental discourse in a global direction.[64] Yet paradoxically, its "global" dimensions depend on the notion of human forcing of change. The local diversity is still evident in the effects, which disproportionately affect the world's poor, who seldom are a major cause of emissions. The "slow violence" of increasing climate change effects already compromises the capacity for many Pacific and "small island state" communities to continue to live in the lands of their cultural traditions.[65]

Albeit unique and somewhat extreme, climate aligns with most other dimensions of environment that we have so far described in this book. It enters into the idea of environment at a certain point in time, adding to its diversity, bringing new expertise, and contributing to the globalizing narrative. In current discourse the phrase "climate and environment" is still often used. This reflects the terms' deeply distinct and long histories and perhaps a distinction between atmosphere and terrestrial land and ocean. It is very rare that someone talks about "demography and environment" or "economic growth and environment" with the same ease. This has nothing to do with climate being less environmentally relevant; rather, it is a sign of the magnitude of the issue and indeed its direct overlap with the environment at its biggest scale. Whereas economic growth, demography, natural resources, public health, and other core areas of the integrative concept of the environment of 1948 still lead their own lives separated from environment, climate change hardly has a life of its own anymore. Natural variability has become a slow background factor. Climate *change* has become purely environmental and scientifically understood as anthropogenic.

"The Earth Is One but the World Is Not"

Our Common Future

As Gro Harlem Brundtland recalled when she was asked to head the UN's World Commission on Environment and Development in 1983, UN Secretary-General Pérez de Cuéllar "presented me with an argument to which there was no convincing rebuttal: No other political leader had become Prime Minister with a background of several years of political struggle, nationally and internationally, as an environment minister."[1] Her own career, as she rose to be the premier of Norway and became widely respected internationally, gave some hope that the environment was not destined to forever remain a side issue in political decision making. In October 1987, after three years of work and consultation, the commission published its report, *Our Common Future.*

The report made the term "sustainable development" famous. The Brundtland report is often seen as a landmark in developing international cooperation in environmental policy. Its formulation of *sustainability*, combined with the stature of a major UN initiative, made that word one of the most influential concepts in late-twentieth-century politics. It formulated the demand "to ensure that [the world] meets the needs of the present without compromising the ability of future generations to meet their own needs."[2]

Four years earlier, in 1983, the commission had been asked "to recommend ways concern for the environment may be translated into greater co-operation among developing countries and between countries at different stages of economic and social development . . . tak[ing] account of the interrelationships between people, resources, environment, and development."[3] Twelve of the twenty-five members of the commission came from the developing countries of the global south, twelve from the north (including two from communist Eastern Europe), and one from Saudi Arabia. The commission differed in this way from much of the scientific and collaborative work undertaken from the time of the Lake Success conferences in 1948, which had been dominated by participants from North America and Europe (see chapter 1). Experts from the industrialized world had identified environmental problems as a global issue, and they had offered their knowledge, skills, and institutional influence as solutions. At last, the politics seemed to be moving away from this approach to reflect something of the global character of the problems. Yet while the report, with its bold assertion of a collective, planetary destiny, may be read as a landmark in united resolve to do something about the environment, it may also be read in another way: as a reflection of the limitations and failings of a global environmental politics that was inaugurated in the late 1940s but that struggled with the diversity and inequity of the world. The environment, conceived and promoted as a tool of conceptual integration, was not delivering as hoped. The planner Lynton Caldwell offered up the concept of the environment in 1963 as a political remedy to the problem of "interlocking crises."[4] But when it came to practical politics, the complexity remained. As the Brundtland report succinctly stated, "The Earth is one but the world is not."[5]

The environmental politics that emerged from the 1960s proved very successful at sweeping up and absorbing many of the long-standing issues such as protecting green space and valued ecological sites or combating pollution. Many of the activists, now called "environmentalists," thought that the transformation had not been enough. For activists, the environment was not a bunch of small issues to be dealt with piecemeal. It was an overarching concept, the most important object of political action. Policy making concentrated on reserving limited, privileged spaces for protection or operated reactively to protect human health. But economic development proceeded much as before.

Rather than integrating policy across the board, the environment had merely succeeded in demarcating its own fiefdom among the competing interests that jostled for attention and money. This was partly because of the way the environment gained a toehold in government, framed by its predominantly scientific expertise and the provision of advice on technical questions. While there had been major realignments within science itself, such experts remained as outsiders looking in on the mainstream preoccupations of economic growth, postwar reconstruction, and geopolitics. This was quite the opposite of what the proponents of the environment had advocated since 1948.

Much of *Our Common Future* dwelled on the lack of communication between the separate spheres of environmental protection and development and with the agencies that worked in each domain. This also reflected further divisions. Environmental policy had largely been instituted at a governmental level in the richer industrialized countries. Leaders of developing countries feared that protecting resources and preventing pollution would impose significant costs on their relatively weaker economies. They argued it was hypocritical to deny the world's poor access to the rich north's levels of income and consumption. Environmental policy was a luxury they could not afford, unless they received significant transfers of wealth to help them.

Even within individual countries, institutions tasked with environmental policy were frequently at odds with those charged with promoting economic growth. One arm of government found its job was to undo the ills bred by the activities of another. Such rifts were different aspects of a more fundamental divide between the desire to allow the economy to unfold in a dynamic fashion and a widespread sense that the resources of the Earth are finite, but to a degree that is hard to anticipate exactly. In other words, the divide reflected the political debates and dilemmas around population and resources that had already been experienced in the industrialized world but that were now given a global dimension (see chapter 3). *Our Common Future* sought to square the circle of optimism about growth with the need for restraint. "Sustainable development impl[ies] . . . not absolute limits but limitations imposed by the present state of technology and social organization . . . by the ability of the biosphere to absorb the effects of human activities." It is hard not to see the biosphere as representing, in the end, an absolute limit (see chapter 7).

Nevertheless, the environment still stood out as an idea that encompassed the whole planet. It defined an expertise that created a new *global* politics. The environment was a truly global issue, which could be scaled to any nation and locality, and no nation could treat it in isolation. Yet the world was certainly not one. Nor was there any regulatory authority that the world as a whole could agree on. Environmental politics has been, in large part, the history of trying to build political institutions that could match the scope and ambition of the concept.

Transnational Science

The idea of the global was fundamental in the moment that the environment emerged. The global postwar moment brought an ethic that humanity *should* unite to protect its vulnerability. In the works of William Vogt and Fairfield Osborn, in the conferences at Lake Success, there was a demand to *do something collaborative* at an international level. It was not obvious how this could be achieved, however, because international institutions were weak and many were only just beginning. Their relationships with national governments remained unclear, and aspiration ran ahead of capacity. Indeed, if anything, this was a time of the unraveling of international connections, as the forced integration of European empires was undone.

Many of the significant early global organizations began with a scientific membership and often with an imperial flavor. This includes the world's first transnational environmental nongovernmental organization, the Society for the Preservation of the Wild Fauna of the Empire, founded in 1903, which still exists as Fauna and Flora International.[6] The International Council of Scientific Unions (ICSU, now the International Council for Science) was founded in 1931 to offer scientific expertise to foster understanding of global issues.[7] Other bodies focused on resources, such as the World Energy Council, originally an offshoot of the World Power Conference held in 1924.[8] The transfer of expertise within empires provided the training ground for many of those involved in postwar development, and indeed the provision of technical expertise to colonies became one of the most frequently presented justifications for empire in its last, gasping phase.[9] Building international networks was relatively easy for those who shared a common research interest—or when the colonized had no say.

The United Nations presented a different rationale and ideal for international collaboration to both empire and the shared norms of science, building collaboration in a world where interests obviously differed. The term *United Nations* was used by Franklin D. Roosevelt in the "Declaration by United Nations" (January 1, 1942). It expanded on the earlier League of Nations and brought with it new styles of governance. Its activities began as an international relief agency in 1943, the United Nations Relief and Rehabilitation Administration (UNRRA), largely dominated by the United States with support from forty-four nations. UNRRA planned, coordinated, and administered "measures for the relief of victims of war in any area" and supplied food, fuel, clothing, and necessities, including medical support services such as vaccinations. Jessica Reinisch has described it as a "tool for internationalizing the New Deal."[10] With the war over but the new challenge of the Cold War looming, the task for the United Nations was to find methods for living at peace in a world where the atomic bomb had made it possible for a single nation to adversely affect all others. Tensions over access to resources were frequently blamed for the war itself: Roosevelt wrote to his secretary of state that "Conservation is a basis of permanent peace."[11] At a time when many nations sought to reconstruct economies shattered by war, the energy and idealism of the United Nations was both a practical support and a source of hope. Those who conceived the United Nations were explicitly concerned about the social responsibility of scientists.[12]

It was not just international governmental initiatives that were important in seizing the postwar moment but also NGOs. The Oxford Committee for Famine Relief began in England in 1942, creating a route for food supplies to be sent through the allied naval blockade to starving women and children in occupied Greece during the war. Later known as Oxfam, it has gone on to provide famine relief, practical education, and reconstruction in Europe and globally. Education became a major focus of the Colombo Plan for Cooperative Economic and Social Development in Asia and the Pacific; it was conceived at the Commonwealth Conference on Foreign Affairs held in Colombo, Ceylon (now Sri Lanka) in January 1950, and again in mid-1951 as a cooperative venture for the economic and social advancement of the peoples of South and Southeast Asia. Its most prominent legacy was funded scholarships in first-world universities for students from developing Commonwealth countries.

In an era when, as we have seen, C. P. Snow declared that it was scientists who "had the future in their bones," science offered international leadership in agriculture, health, medicine, and management.[13] Paradoxically, the scientific community that had created the knowledge for nuclear technologies and was deeply involved in research on behalf of the "military-industrial complex" was often at the forefront of expertise and leadership in this search for an international peace. Scientific cooperation also represented an apparently neutral place for accommodating many ideological hues. As well as already being embedded in international networks, many scientists had worked closely with government as part of the war effort and enjoyed a massive boost to research support in the postwar era and in the Cold War. The run-up to war, and even more so the conflict itself, had led to the foundation of a range of funding bodies that would be fundamental in steering postwar research, such as the Office of Naval Research and the Atomic Energy Commission in the United States and the Centre National de la Recherche Scientifique (CNRS) in France.[14] Teamwork and application became highly valued, which also contributed to a reframing of the "imperial mission" as one of expert-led development after the war.[15]

Much of the huge increase in research and development (R&D) was focused on weapons or means of war, which included weather modification and insecticides, insects being the main cause of casualties among allied troops in the Pacific theater.[16] In the early Cold War, enormous amounts of R&D funding in the United States came from the federal government, and some 83 percent of that was devoted to defense. This was a "permanent mobilisation of science" for national security, in Jon Agar's words, but it also had significant civilian spin-offs.[17] Cybernetics and modern computing had their immediate origins in projects at the Massachusetts Institute of Technology dedicated to improving the accuracy of antiaircraft fire and flight simulators.[18] Defense interests made huge investments in meteorology and climate science, and much of this knowledge was disseminated through international networks (see chapter 5). Both conceptually and practically these approaches and technology would have enormous influence on understandings of the environment, whether in shaping parts of ecological science in the second half of the twentieth century or with the emergence of global models for the Earth system. Only a comparatively small number of scientists expressed environmental concerns amid this concerted effort.

During the Cold War, scientific collaboration was seen as a means of international rapprochement and a way to exchange useful knowledge between competing powers. This was signaled in the International Geophysical Year of 1957–58, and such initiatives took institutional form with the International Institute for Applied Systems Analysis (IIASA) set up in 1972 near Vienna. Cybernetics was viewed as a route to collaborate around common concerns of managing complex industrial societies, evading ideological divisions. As a novel kind of metaspecialization, it could also cultivate innovative modes of interaction and avoid old disciplinary hierarchies, or what the (critical) sociologist Isabel Hoos at the time called "expertness."[19]

Bodies to accumulate and disseminate technical expertise were also set up directly under the auspices of the United Nations. Often these initiatives built on and transformed earlier transnational networks of science. The UN Food and Agriculture Organization (FAO) was conceived under the tutelage of Roosevelt in another of those luxury resorts beloved of international meetings, tucked into the Allegheny Mountains in Virginia. Formally founded in Quebec City in October 1945, the FAO was a means for dealing with Malthusian concerns about population outstripping resources, through working to develop technical assistance and food aid. In 1948 it took over the functions of the older International Institute for Agriculture and eventually moved to that organization's base in Rome. The FAO's first director, John Boyd Orr, had been active around issues of food, malnutrition, and population since the 1930s and published *Food: The Foundation of World Unity* in 1948.[20]

Initially an educational and cultural organization had been proposed for the United Nations, but by November 1945 it became UNESCO—the "S" for "Scientific" was added in recognition of the importance of science to global peacemaking.[21] Instrumental in adding the S was its first director-general, biologist Julian Huxley. He could hardly have been better connected in the elite and cosmopolitan circles of transnational science. His grandfather was Thomas Huxley, friend and advocate of Charles Darwin, and he was related to the English poet and essayist Matthew Arnold, among other writers and editors. One brother, Aldous, was a famous novelist, and another brother, Andrew, later won the 1963 Nobel Prize for Physiology or Medicine. In the 1930s, Julian Huxley made a significant contribution to the development of the "modern synthesis" of evolutionary theory and wrote extensively about ecology (see chapter 4). He worked in Germany, in the United States, and for the Colonial Office in east Africa and saw varied service in both world wars. Huxley was international, collaborative, at ease in both arts and sciences. Together with the biochemist and Sinologist Joseph Needham, the first head of the natural sciences section of UNESCO, Huxley had been a leading light in an earlier interwar movement for "social responsibility of science." Needham spent much of the war running a Sino-British office for scientific cooperation in the wartime nationalist capital of Chongqing. Together Needham and Huxley saw unique possibilities for international and "universal" outreach through a scientific secretariat within the United Nations.[22]

At this time a range of organizations both old and new (including private endowments such as the Rockefeller, the Wenner-Gren, and the Ford Foundations) began to coordinate discussion and action on conservation, climate, food security, resource scarcity, and family planning.[23] Crucial characteristics of this broad trend were the development of *teams of experts*, usually drawn from scientific disciplines, to pronounce on problems of a planetary scale. Equally, they both answered and generated an expectation that something had to be done about the problems by those charged with governance—indeed, creating new systems of governance where necessary. Somehow, these had to embrace both a planetary and a local scale.[24]

Resources, conservation, and eventually the global environment demanded technical and scientific management as well as the moral responsibility of stewardship. Once the environment was conceived of as global and integrated, there could be little doubt that environmental management needed to operate on a planetary scale. The environment comprised more than just the lands of nations and the territorial or even international waters of the sea. The global atmosphere was also part of its purview. Borders between territories were irrelevant to the circulating atmospheric system and deep ocean currents—and in turn the pollution that traveled via these means. In the longer period from the 1940s to the 1990s, as the environment was conceptualized and globalized, it became a contributing factor in moves toward "globalization" in governance and policy alongside development, trade, and keeping the peace. These challenges would pose new (and in some cases, as yet intractable) problems of governance beyond the capacities of legal and administrative systems still largely cocooned within nations.

Conservation on a Mission

The challenges of negotiating a way between old and new approaches and dealing with the environment as a phenomenon across many scales are exemplified by postwar international conservation. UNESCO convened the conference that created the International Union for the Protection of Nature (IUPN; later IUCN, substituting "Conservation" for "Protection"). This was in many ways a revival of prewar conservation networks, and it had been controversial whether, as some wanted, a new body should fall under the auspices of the United Nations or be an independent body. Having UNESCO convene but not run the IUPN was a compromise, but one that left the new body short on funding. At this stage, leaders in the organization were, in the words of British ecologist Max Nicholson, "emotionally inspired missionary individuals," and participants included William Vogt and Fairfield Osborn.[25] Money was largely spent on projects in the industrialized north, while IUCN conferences in the period from 1948 until the 1960s were particularly concerned with African wildlife. This strongly reflected imperial connections and a focus on the "charismatic megafauna" loved by big game hunters. South Africa emerged as a major player in these ventures with its tradition of big game hunting (and protection), until after 1961, when the apartheid state's increasing isolation led it to withdraw from international initiatives. The desire to protect species and also spectacular sites like Murchison Falls on the White Nile in Uganda led to tensions with independence movements and the leaders of new states wanting to build up their infrastructure and hydropower.[26] Although the IUCN sought to stress the tourist potential of wildlife, the FAO remained unconvinced that the conservation movement took the need to use resources for development seriously enough.[27]

It was in part the lack of funding for the IUCN that led in 1961 to the foundation of an NGO, the World Wildlife Fund (WWF), through an agreement (the Morges Manifesto) hammered out on the shores of Lake Geneva. Signatories included many prominent conservationists, including—among the many other things we have seen—African wildlife enthusiast Sir Julian Huxley, IUCN vice president and ornithologist Sir Peter Scott, and director-general of the British Nature Conservancy, another ornithologist, E. M. (Max) Nicholson. Nicholson had also been at the Princeton Inn to assess man's role in changing the face of the Earth on that balmy June day in 1955 and later attended the Future Environments of North America conference in Colorado in 1965 (see chapter 2). The organization's president was Prince Bernhard of the Netherlands, and it benefited from the strong and active endorsement of the Duke of Edinburgh, important in the countries of the British Commonwealth. It was in this capacity that the monarch's husband provided a preface to the British edition of Rachel Carson's *Silent Spring* alongside Julian Huxley. One of the new organization's first fund-raising ventures was to organize safaris in East Africa under the headline "Making African Wildlife Pay," with the double meaning intentional.[28]

The WWF continued an older conservationist vision of preserving wild places without humans, designating areas from which human disturbances should be excluded. This was strongly influenced by the National Parks movement, particularly in its American form. Touting "America's Best Idea," national parks activists had set aside large areas of protected habitat that had been managed under government supervision since the 1860s. National parks were part of a concern for "wilderness"—nature without humans—that became part of American nationalism in this period and later became an important part of environmentalism.[29] Leaders in these ideas were John Muir, founder of the Sierra Club, and later ecologist-forester Aldo Leopold.[30] In 1962, the United States hosted the first World Congress of National Parks in Seattle, and two years later the US Congress passed the Wilderness Act (1964).[31]

However, efforts to define a *science* for choosing places to reserve as national parks proved difficult.[32] Areas had often been chosen for their aesthetic value or strong associations with national identity. In 1948 the IUCN sponsored a World Commission on Protected Areas to facilitate "the establishment of national parks, nature reserves and monuments and wild life refuges, with special regard to the preservation of species threatened with extinction."[33] The Australian Academy of Science, responding to this IUCN directive and confronting very different systems of nature management in Australia's different states, tried to establish ecological (scientific) principles to inform the choice of places to be reserved for nature. The Academy recommended "gap analysis"—choosing land parcels that would preserve as many different representative ecosystems as possible.[34] While the idea of representative ecosystems is more based on "trust in numbers" and certainly rather different from the aesthetic of wilderness, the expansion of national parks that carried on through the 1970s was as much because of the rise of political support for environmental causes as it was evidence of using scientific data for policy making.[35]

Densely populated Britain saw nature conservation focused on protecting animals and plants rather than reserves. Arthur Tansley, founder of the ecosystem concept (see chapter 4), became foundation chairman of the newly founded (British) Nature Conservancy in 1949, working in partnership with fellow ecologists and elder statesmen Charles Elton, Max Nicholson, and Dudley Stamp. The conservancy's mission was to provide good science to inform wildlife management and nature conservation (see chapter 4).[36] This worked parallel to but largely independent from national parks legislation that was drafted, not by ecologists as it was in the United States, but by architect and planner John Dower.[37] Faced with limited funds and the need to compensate landowners for restrictions on use, the Nature Conservancy concentrated on designating and protecting small sites associated with valued species.

Thus, in many countries the most renowned sites of conservation were reserved, nationally valued landscapes thought to evoke something of the nation's spirit and were often popular among relatively wealthy hunters and hikers.[38] This form of preservation was often closely aligned with the protection of historic monuments, and by the mid-1960s they became linked in the idea of "world heritage." Nature conservation and cultural conservation came together with the International Council on Monuments and Sites (ICOMOS), founded in 1965. ICOMOS expanded the original definition of *heritage* from conserving buildings (from the Athens charter of 1931) to conserving whole sites, increasingly including sites valued for "natural heritage." The World Heritage Convention, adopted by the General Conference of UNESCO on November 16, 1972, linked together in a single document the concepts of nature conservation and the preservation of cultural properties, recognizing the value of people interacting with nature. Just as "wilderness" had become a nationalist value for Americans, so certain sites of heritage and nature became the property of all humankind. These values were complementary rather than antithetical. In that same year of 1972, the US Congress supported a lavish celebration of the "centennial of National Parks," one hundred years after the foundation of Yellowstone National Park.[39]

National parks had become an international movement, and the postwar decades saw the consolidation of a number of NGOs working globally for conservation. Their practice, however, had largely been to isolate and manage treasured landscapes, often with little reference to local populations. Equally, this scale of ambition seemed very far from the planetary impact of the environmental problem catalogue (see chapter 1). Nature preserves, including national parks, areas of scientific significance, and state forests, constituted only a minor alleviation of pressures on the environment and did next to nothing to deal with the systemic aspects of economic growth that were generated. At the same time, they often represented an unwelcome intervention to people and governments in the developing world who experienced conservation as a continuation of imperial condescension. It was precisely in the desire to overcome these suspicions that an IUCN report drafted by a new generation of leaders of the IUCN and published as the "World Conservation Strategy" in 1978 would coin the term "sustainable development."[40]

Aggregating Expertise

While biologists were prominent in the IUCN and the WWF, other disciplines were busy building international networks and interacting with politicians and policy makers. By the early 1960s people were beginning to hear the aggregative term *environmental sciences*. It was largely in this decade that the aggregation of environmental expertise moved beyond the first gatherings of dozens of experts into large transnational teams working on an ongoing basis. In doing so, they created new modes of work, harnessed new computer technology, and subjected the world to a veritable deluge of acronyms to describe their organizations. Increasingly, they moved beyond dealing with scientific problems and data gathering per se to the challenges of *governing the human environment*, as it was increasingly termed by the late 1960s. Nevertheless they remained reliant, in the

last instance, on the traditional means of creating change: the acqui-
escence of national politicians and the provision of money.

The geosciences grew internationally in the late 1950s, particularly through the great strategic and scientific success of the International Geophysical Year (IGY 1957–58; see chapter 5). Despite the
Cold War, sixty-seven countries from both East and West collabo-
rated and participated in IGY projects; it helped provide momentum
for the Antarctic Treaty of 1959, which "froze" territorial claims from
a variety of nations to the southern continent but also fixed it as a
place for shared scientific research, permitting this as a legitimate
reason for newcomers to set up stations irrespective of others' presence, and banning resource extraction and military activity. This was
the very moment when the space age began with the Soviet Union's
Sputnik 1 launched on October 4, 1957. The Antarctic Treaty later
provided a model for the regulation of outer space. The importance
of IGY to the environment was that it was used as a template for organizing large-scale international collaborative research.[41]

The logic of global problems was seen to point toward global institutions and aggregations of experts. Sometimes this led to the
augmentation of the activities of enduring bodies. The FAO would
provide the base for a Committee on Pesticides in Agriculture following concerns raised most prominently by Rachel Carson's 1962 book,
Silent Spring.[42] By December 1970, its remit was extending beyond
the land to consider marine environments, sponsoring a technical
conference on marine pollution and its effects on living resources
and fishing, and linked with growing scientific concern about the
health of coral reefs.[43]

Climate and weather was another major focus for global effort,
albeit not yet considered to be "environmental" (see chapter 5). Information from any one part of the world helped with forecasting
elsewhere, and traditions of international cooperation had been established in the nineteenth century.[44] The World Meteorological Organization (WMO) was established in 1950 and a year later became
the specialized agency of the United Nations for meteorology (weather
and climate), operational hydrology, and related geophysical sci-

ences.[45] It was, together with ICSU, the chief organizing partner of the IGY program. The late 1960s and early 1970s saw major droughts in the American Midwest, in Russia, in Africa, and in Australia, which raised wider concerns about possible global climate change.

The traditional sites of power and influence in the West continued to play a major role. Under the leadership of Cambridge systems biologist C. H. Waddington, the International Biological Programme (IBP) was set up to coordinate international Big Ecology, a suite of large-scale projects funded by national governments and groups of governments. Inspired by the IGY and its opening conference in Paris run by the ICSU, the IBP developed a global view on ecosystem ecology and complex environmental issues as it explored "The Biological Basis of Productivity and Human Welfare." The IBP's progress was reviewed in 1968 by the UNESCO Conference on the Rational Use and Conservation of the Resources of the Biosphere (known as "the Biosphere Conference"), which in turn proposed the ongoing Man and the Biosphere Programme (MAB). This group, established in Paris in 1971 and still in operation, aims to use science to "improve relations" between people and environments. Its chief mode of operation is to propose, investigate, and instigate members of the World Network of Biosphere Reserves. In 2013, there were 621 biosphere reserves in 117 countries.[46]

In late 1968 a Swedish proposal went before the UN General Assembly, proposing that the organization sponsor a major Conference on the Human Environment to be held in Stockholm. Sweden, with its deep involvement in transnational science, tradition of neutrality and internationalism, and strong domestic culture of outdoor life and conservation, saw itself as well placed to undertake a catalytic role. The passing of this, Resolution 2398, would trigger a wave of efforts to coordinate, build, and deploy expertise. Another ICSU initiative, started in 1969 as an ad hoc committee and becoming known as the Special Committee on Problems of the Environment (SCOPE, later Scientific Committee on Problems of the Environment), was instrumental in preparations for the Stockholm meeting. Simultaneously, crucial evidence was being gathered in support of the hypothesis of global warming caused by carbon dioxide concentration in the atmosphere and had already appeared in the appendix of the report on pollution delivered to President Lyndon Johnson in 1965. Global warming was an issue in the MIT project of 1970, "Man's Impact on the Global Environment," which concluded conservatively that global warming might contribute to "widespread droughts, changes in the ocean level and so forth" (see chapter 5).[47] The MIT findings, almost exclusively involving US scientists, led to a second international meeting in Stockholm in 1971, a Study of Man's Impact on Climate (SMIC), with fourteen nations represented. SMIC's warnings of melting polar ice caps, diminishing albedo, and potentially drastic impacts on climate and environment became important background reading for the delegates to the 1972 conference. Influential and networked individuals continued to play a major role. The Royal Swedish Academy of Science (KVA) launched *Ambio*, a journal of environmental research, at the time of the 1972 conference. The Boston and the Stockholm SMIC initiatives were organized by MIT professor Carrol Wilson, who had cut his teeth as assistant to Vannevar Bush, impresario of wartime research and development in the United States. Wilson later worked in the Atomic Energy Commission and as a UN advisor. He was a member of the Club of Rome (see chapter 3) and had a hand in promoting the *Limits to Growth* report, all of which were timed to coincide with the lead-up to the Stockholm conference in 1972.[48]

In the run-up to Stockholm, there was an overwhelming centrality of scientific expertise in developing international environmental politics, providing both the means to conceptualize the environment and the authority to justify interventions. The environment was, to a major extent, defined through these processes. Diplomacy, policy, and, indeed, the environment, were created in tandem (or "co-created," as historians of science like to put it) through the formation of an internationally active and restlessly conferencing alliance of scientists. The environment was, by these means, in many ways a creation of science and especially science, as scientists sought to shape institutions that could influence policy and arrest environmental destruction. The environment was envisioned and imagined through the activities and outputs that these networks produced—imagined as a kind of networked planet.

It has been a commonplace to perceive the historical development of the environment as almost exactly the reverse: *first,* the omnipresent environment "out there," the critical state of which was revealed by a few vanguard scientists and alarmists, *then* the global organizations and policy units, previously existing or newly invented, that mobilized to act on the issue in its various manifestations, from overpopulation to soil erosion to biological diversity to climate change to waste management. This is not the only way to portray this period of profound and revolutionary change in the relationship between humanity and its planet.

The major outcome of the Stockholm conference was the foundation of the United Nations Environment Program (UNEP), a body that could coordinate networks of expertise and channel funding toward environmental policy. It would be based in Nairobi, because of a late intervention from governments of developing countries that remained skeptical about policies and arguments that largely originated in the well-funded academies of industrialized nations. From 1972, UNEP drew heavily on the established networks of aggregated scientific expertise in creating its complex definition of "the world environmental situation." The UNEP website today is organized in six core areas: Climate Change, Disasters and Conflicts, Ecosystem Management, Environmental Governance, Harmful Substances, and Resource Efficiency.[49] All the elements of today's six "core areas" had emerged in various agencies inside and outside the United Nations from the late 1940s and 1950s. They were not an initiative of more recent environmental politics or post-1960s environmentalism.

The prominence of science and its organizations in understanding the environment was never a "given." It was the culmination of many different efforts, which is why we pay so much attention to it in this volume. This is easier to do with hindsight than when it was happening. A counterfactual history of the environment would, for example, have linked precisely those issues that came to be "the environment"—pollution, poverty, epidemics, threats to public health, devastated soils and livelihoods, depletion of fish and game—not to the sciences but instead to social justice, civic rights, animal welfare, form and aesthetics, or urban planning. These concerns were certainly present and are well documented, especially in accounts of local and popular environmentalist campaigns, but were rather marginal in the environment's international framing in the 1970s.

We continue to live in a world of acronyms and aggregated expertise. Now it is the group Future Earth that gathers together the partnerships of ICSU, such as the former Earth System Science Partnership, the former International Geosphere-Biosphere Programme, the former DIVERSITAS (an international program of biodiversity science based in Paris, France), and the former International Human Dimension Programme on Global Environmental Change (IHDP), in Bonn, Germany, along with over twenty continuing projects.[50] Future Earth seeks to deal with not only science but also all the dimensions of environmental change, human and biophysical, the "planetary boundaries" for a "safe operating space" for humanity on Earth, providing information to the IPCC and the World Climate Research Programme (WCRP) in Geneva, Switzerland.[51] One might reasonably feel that it requires special training simply to keep up with the proliferating networks and names of researchers and their interconnections.

Although the years since 1972 did not fulfill all the goals the conference had hoped for, the development of an international capacity for environmental governance and harnessing networks of expertise has yielded notable successes. The Montreal Protocol on Substances that Deplete the Ozone Layer (signed in 1987) agreed on regulations for reducing the production of damaging chlorofluorocarbons (CFCs), with an ongoing monitoring system with numerous revisions to the original protocol and a mechanism for compensating poorer countries for the costs of enforcement. It was the first UN treaty to be ratified by every single member of the organization and has been very successful in its environmental results. In 1988, UNEP and the WMO joined forces in founding what is perhaps now the most influential and certainly best known of such integrated networks: the

Intergovernmental Panel on Climate Change (IPCC). This occurred at what has been called, tellingly, the "Woodstock" of carbon dioxide, held in Toronto.[52] Yet the perhaps self-validating reference to 1960s counterculture and generational revolution obscures the real origins of this approach to environmental politics and its institutionalization. The year 1988 was, not coincidentally, a very warm one in planetary terms. Climate scientist James Hansen announced, "the greenhouse effect is here" and gave electrifying testimony before the US Senate on a hot and sultry day, as forest fires ravaged Yellowstone Park.[53] Global warming was well known and long established among scientists (see chapter 5), but with the IPCC that collective authority could be swiftly translated into an institutional response in the halls of power.

Think Global, Act National

In the twenty-first century, global environmental governance in UNEP also draws on law, diplomacy, and policy to guide "national, regional and global" thinking through its Environment Management Group (EMG), a UN "system-wide coordination body on environment and human settlements."[54] But in the end, policy has to be enacted to have effect: it requires a legal framework. And for all the integrative, global imperatives of the environment, it has remained the case that most of the implementation of environmental policy takes place through the medium of nation-states, while international action requires the collective assent of national governments. On the one hand, this represented an avenue for success. Very few countries do not now have significant bodies of officials tasked with environmental protection. On the other hand, true revolutions in government are rare indeed. The new wine has frequently been poured into old bottles, replicating a division of labor and expectation that those who first promoted environmental policy had hoped to overcome. These problems were aggravated when the primary coordinating bodies between national governments remained conferences of national leaders, diplomats, and ministers with many other responsibilities and interests. Countries persisted in seeing their own interests as divergent.

Both within and between countries, the main dividing line was often drawn between those who focused on economic growth or environmental protection and saw the other as a barrier to their ambitions.

When the United States' President Johnson received the major report *Restoring the Quality of Our Environment* from his Scientific Advisory Committee in 1965, he did not comment on the section on climate (see chapter 5). He commended the panel on its thorough investigation of pollution and noted progress to a "cleaner world" in legislative actions by the 89th Congress, noting the Water Quality Act of 1965, amendments to the Clean Air Act, and the Highway Beautification Act. He then drew attention to the "more than 100 recommendations in the report" and asked the appropriate departments and agencies to consider the recommendations.[55]

The environment, like government, runs on many fronts rather than just one, and the emerging legislation dealt with quite traditional concerns in strictly demarcated ways. Yet by the 1960s a large body of new experts was needed to keep pace with measuring and responding to environmental hazards, as traditional approaches had become inadequate to the scale of the task. On January 1, 1970, President Richard Nixon signed the National Environmental Policy Act into US law, establishing the Environmental Protection Agency and requiring environmental impact statements as part of the planning process. Lynton Caldwell, who had first called for environmental policy just seven years before, was instrumental in its drafting. Both the style of institution and the policies it espoused were followed rapidly around the developed world.[56]

The emergence of national environmental governance had an essential transnational dimension. The American case provided a direct inspiration in West Germany, for example, where the more traditional conservation (*Naturschutz*) began to be supplanted by "environmental protection" (*Umweltschutz*) in 1969, a concept dreamed up on November 7, 1969, at the behest of officials in the Ministry for the Interior, where environmental causes were taken up by politicians from the junior governing coalition partner, the liberal Free Democratic Party.[57] This led to a fairly typical interdepartmental tussle between the Ministry of the Interior and the Ministry of Agriculture for authority, in which the Interior Ministry won out, taking the lead role in Germany's new Environment Program from September 1971 onward. The program coordinated a wide range of expert consultation and advice that had been built up since the time of a major review of how development was affecting the "household of nature" in 1966, leading to fifty acts of legislation by 1976.[58] An enduring legacy of such action was that government bodies were slotted into a hierarchy of ministerial power, with policies often assessed as a trade-off between environmental protection and the more powerful bodies tasked with economic development.[59]

In Britain as in West Germany and the United States, environmental policy rose on a wave of support across political parties, being put firmly on the agenda with a group in the Cabinet Office coordinating policy at the heart of government by Labour Prime Minister Harold Wilson in 1969 and then elevated to a new Ministry of the Environment by the incoming Conservatives in 1970, the same year that a ministry was established in France.[60] The new politics was swiftly taken up at the level of the European Community (EC) too, which inaugurated its own Environmental Action Programme by 1973. Being environmental was a way to look modern, to seem responsive to public concerns, to be an international leader. There were, of course, very practical worries and concerns, many of which received new legitimacy from the international mood. The EC quickly recognized that national legislation had implications for the operation of a free internal market, necessitating the harmonizing of regulations. As the EC expanded over the following decades to become the European Union (EU), directives and laws became an important vehicle for raising environmental standards throughout the nations of the organization. Through *Natura 2000*, new member states were required to meet environmental standards before they were allowed to join the EU.

The management of environmental protection was not just something confined to the West. By 1985, more than 140 nations had environmental protection agencies of different sorts.[61] India had a long tradition of localized movements campaigning on what came to be called environmental issues, the most internationally famous being the Chipko movement, discussed below. The national government responded to the impetus created by the UN's 1972 conference in Stockholm, resulting in India's National Committee on Environment and Planning being established that year, followed by statutes dealing with water pollution (1974), air pollution (1980), and environment (1986). Amendments of 1976 and 1993 enshrined the environment in India's constitution, declaring, "the state shall endeavour to protect and improve the environment and safeguard the forests and the wildlife of the country."[62] China, which had sent delegates to Stockholm, engaged formally with environment policy (alongside general reform) from 1979. China's National Environmental Protection Agency was created in 1988.[63] However, in both the cases of China and India, central environmental agencies employ relatively fewer people than the US's Environmental Protection Agency, and much of the power of implementation rests with state and provincial authorities, making for very uneven rule enforcement; this may especially be the case where the same local authorities are tasked with generating economic growth. The emergence of an explicitly environmental governance structure at the national level was only one aspect of institution building within these nations. Environmental governance structures in turn generated new political debates about participation and relations with relatively marginal groups, raising questions of expertise, debates also part of environmental issues in international institutions.[64]

Most of this new national government activity subsumed older forms of environmental management, of which there were two major traditions. One was the regulation of water and air pollution, which had a heritage in municipal government stretching back to medieval times. In eighteenth-century Paris such regulation was already being passed to technical experts who negotiated the permitted parameters with industrial interests. Over time, with the vastly increased output of smoke and chemicals associated with the Industrial Revolution, such expert cadres emerged across western Europe and North America as well as in some colonial cities between the 1850s and 1950s.[65] When individual instances of pollution became problematic it became habitual to convene commissions of experts to propose remedies, and thus when pollution was absorbed into a more general category of environmental problems in the 1960s, these experts swiftly became the core staff of new regulatory bodies. The other tradition was nature conservation and landscape protection (see chapter 4 and above). This was often allied with aesthetic ideals and antimodernist sentiment seeking to restrict development in treasured landscapes as well as efforts toward species protection. From the late 1960s the very many local and national bodies associated with the "amenity" of a pleasant landscape and nature protection found themselves part of a broader "environmental movement," often with some reluctance and a sense of their true interests being overshadowed.[66]

Despite the formal integration into a single policy area, in practice such strands often, frustratingly, still work independently. This did not preclude considerable success in enacting some policy. The extent of the landscape granted some kind of formal protection was greatly extended, both nationally and globally, although with little impact on continued biodiversity loss. Equally, the emissions of some targeted chemical pollutants fell dramatically during the 1970s, water quality improved, and smoke pollution abated. National environmental politics thus cemented the presence of scientific expertise in government and these experts as arbiters of what was good for the populace. It often did so, however, drawing on traditions that had been quite cautious and consensual, where planning and chemical controls in agriculture worked closely with landowners and the monitoring and control of air pollution drew on the engineers who staffed industrial enterprises. These practices were quite different from the more dramatic demand for change from globally oriented critiques of development and environmental destruction. But both the global critique and the culture of national regulation tended to treat environmental problems as depoliticized matters of fact. On a national level, as was the case with transnational networks of science, environmental problems were often portrayed as merely technical problems amenable to resolution by science. In the eyes of politicians this might be their prime virtue, allowing action that seemed free from ideological positioning and that could appeal "across the aisle" to suspicious voters. Over time, this value-free presentation of environmental problems, which had always been subject to some criticism, became increasingly difficult to sustain.

Government by the Nongovernment Sector and the New Geopolitics

Government does not only take place at the national level or indeed in parliaments or executive bodies and agencies. Communities, firms, and individuals have governed the environment and sought to maintain their livelihoods for centuries, indeed millennia. We can identify struggles over the control and management of resources all over the world for many centuries. The Chipko movement of "tree huggers," for example, from the Himalayan foothills, contested the right of loggers to extract timber from their community woods in 1973 by putting their bodies in the way of the loggers, "sticking to" the trees (Chipko means "to stick"). Such struggles between communal use and commercial extraction have been familiar in Asia and Europe since medieval times. In these cases and from the perspective of those whose livelihoods are affected, what is new to our age of environment is the scale and scope of resource extraction in a globalized economy. The conflict is age old. The environmental movement might be seen as a quintessential social movement of the 1960s in the developed world, mobilizing huge numbers of people into actions such as Earth Day (April 22, 1970), whose gathering of twenty million participants in the United States is sometimes described as the biggest political mobilization on a single day. In reality such popular mobilization had a very long—if more geographically restricted—history. Arguably what was novel about such protests by the 1970s, whether against modern forestry and timber cutting in India or even involving the descendants of the "conservation movement" founded in the United States in the early twentieth century, was that they all were seen as part of collective environmental politics.[67]

And so the defense of livelihoods and sometimes sacred connections with land, often organized through village-level institutions, has been reinterpreted as the "environmentalism of the poor."[68] Nevertheless, as we have seen, environment has also been represented as a luxury good, as something to be enjoyed only when more fundamental goals of poverty eradication and development are achieved. This reflects the dominance of the idea of the environment that had emerged in the prosperous West; it was an expression of values and standards among networks of expensively educated, rather technocratic experts—at least as they appeared officially in their institutions and conferences—because the values and ethics of individual scientists were not, of course, standardized or indifferent to the plight of others. The alleged trade-off between development and environment was the very thing that the Brundtland report sought to overcome, with a commission drawing widely on knowledge from the global South.[69] Yet the cleft between a science that spoke for nature and arguments for growth that allegedly spoke for the concerns of ordinary people has been persistently deployed in environmental politics. Who actually reflects the views of whom is another matter altogether! In truth, the emerging environmental politics often dramatized political fault lines and questions of authority, ownership, and rights that were very well established. And for all the claims of scientific bodies to present information, limits, and solutions in a strictly neutral mode, in the end environmental issues are still being associated with parties and ideologies on the established political spectrum.

These divides were already stark at Stockholm in 1972, often articulated by national governments as well as social movements. In the run-up to that Conference on the Human Environment, representatives of the military dictatorship in Brazil brought objections to the United Nations that environmental concern was a "diversionary manoeuvre by the major aid donors." India's premier, Indira Gandhi, garnered standing ovations with her address at the end of the conference proclaiming that poverty, not pollution, was the main global problem.[70] Developing countries had fought hard in the postwar decades to assert national sovereignty over their own resources, and they did not want to become subject to international regulation in the name of the environment. This insistence on sovereignty meant that environmental governance was, in the end, subsumed to national concerns and framed as part of development, a conception that would come to dominate the agenda and reflected in the mantra "sustainable development" found in the Brundtland report.[71]

As leaders of the G77 group of developing nations, Brazil and India led demands for "additionality"; that is, if the cost of development was increased by environmental policies prioritized by the rich world, then developing countries were owed transfers of money for the greater expense of development than countries who had become wealthy at an earlier date without such constraints. The resolute refusal of the United States to countenance such transfers in both 1972 and at later meetings scuppered any hopes of more ambitious coordinated policies at the international level. Instead, UNEP would operate more modestly on a project-by-project basis.[72] Brazil and India would be centers of controversy during the 1980s, when loans from the World Bank for major highway and dam projects became targets of campaigns by environmental NGOs arguing that their ecological impact was not being sufficiently taken into account. A newly democratic Brazil took the helm at the 1992 Earth Summit held in Rio de Janeiro, which marked and built on the two decades' of work since Stockholm, but major agreement was blocked by very similar issues as in 1972. The deliberate foregrounding of the global South's need for financial compensation failed to win over a US administration increasingly wedded to free market solutions to political problems and suspicious of any imposition of economic or ecological limits (see chapter 3).[73]

The difficulty of reaching political consensus at an international level, as well as the inequalities and disagreements within countries, has meant that an international environmental politics that might aspire to match the global vision of the environment has been pursued largely by NGOs. With the lack of intergovernmental consensus at the 1992 Rio conference, the main outcome was the signing of Agenda 21, a nonbinding agreement promoting action to foster sustainability at more local levels with strong NGO participation. NGOs operate transnationally yet often manage specific local projects within countries. The desire for an integrated approach to problems, evoking the interconnection between the experience of people in a single slum, village, suburb, or common land and national or global phenomena, is not easy to match with practical means to give voice to all the interested and often wildly unequal parties.

There are many excellent histories of these struggles and of contests around environmentalism. It is not the purpose of this book to retell those more familiar stories, but they are important, and so we highlight a few here. We have already noted the growth of transnational NGOs such as the World Wildlife Fund set up in 1961, Oxfam (1942), and Conservation International (1987). More explicitly, campaigning organizations such as Friends of the Earth (an antinuclear offshoot of the American Sierra Club in 1969 that became an international organization in 1971) have become global campaigners of influence with mass memberships but are concentrated again in the industrialized world. NGOs provide accreditation of sustainable fisheries, forestry, and organic agriculture to encourage environmentally conscious supply chains and allow consumers to identify less damaging goods. A striking phenomenon is the extraordinary range of activities that are now considered to be part of environmental politics, from the community-level management of forest resources in India studied by Arun Agrawal to the many campaigning groups of varied social backgrounds that helped secure the green belt and combat toxic pollution in the San Francisco Bay area, as told by Richard Walker.[74] One could fill books simply by listing them. Urban lobbying groups, operating at the scale of a city, have become increasingly prominent in recent years. Cities are taking global environmental thinking in new directions, perhaps especially in the nations sometimes referred to as the BRICS (Brazil, Russia, India, China, and South Africa).[75] China's exceptional city-growth since the 1980s, Brazil's shantytowns of risk and poverty, and South Africa's emphasis on "brown" environmental concerns since the end of apartheid in 1994 are all reasons to consider cities the focus for new-style environmental revolutions in the twenty-first century.[76]

Campaigns have not been limited to the land or atmosphere. Lobbying for the rights of whales and other cetaceans was central to the identity of the NGO Greenpeace, first established in 1971. One of its prominent victories was the 1989 UN moratorium on large-scale driftnets. Greenpeace has been involved in a wide variety of campaigns since—against toxins, pesticides, and nuclear testing, among others.[77] Environmental organizations founded to protect particular landscapes or species increasingly felt compelled to take the integrative path of addressing a wider set of the environmental problem catalogue, of linking the local and global, of demanding more coordinated policy and thinking, and sometimes developing their own networks of aggregated expertise. They are, in their way, a legacy of 1948.

The environment is not unique in creating increasing and novel demands on global governance. Global governance and regulation often concerns areas beyond nation-states, as can be found in the law of the sea or aviation airspace agreements that enable international traffic to move around in an orderly fashion and in the rules about international carriage of letters as managed by the Universal Postal Union. Human rights and security questions often have global implications and regulatory bodies to ensure parties respect agreements. Biophysical systems work across arbitrary national boundaries such as air space, fishing zone rights, and river watersheds. When acid rain from industries in the United States falls in Canadian forests or sulphur dioxide pollution blows across Europe, international law, transnational negotiations, and regulations become involved. Indeed, most environmental agreements since the 1960s have been multilateral treaties among neighbors regarding specific cross-boundary issues.[78]

Winds and currents flow with no respect for borders. In preparation for the UN Stockholm 1972 conference, the US government issued a booklet entitled *Only One Earth* in several languages, stating: "For pollution and ecological degradation national borders do not count." A growing sense of limits since the 1940s added to this planetary perspective. This is where the environment, an entity that covers the whole globe but that is also fragile and in need of management, shapes the possibilities for governance beyond national interests. It

Seeking a Safe Future

A Safe Operating Space for Humanity

In a 2009 paper in *Nature*, Johan Rockström of the Stockholm Resilience Centre and a team of colleagues considered a range of biophysical indicators of planetary health specifically to define a "safe operating space for humanity with respect to the Earth system."[1] They noted the long stable era of the Holocene over the past ten thousand years that had accompanied the shift of human societies away from hunter-gathering, followed by the rapid shift in the planet's biophysical subsystems and processes in response to the rise in fossil fuel use by humans, largely in the past century. While people have always been dependent on the physical environment for clean air and water, for food and fuel and so forth, they argued (as had others for many years) that the economic system had assumed these things to be infinite and had discounted their value. The concept of "ecological services"—an idea whereby a monetary value could be attached to the services of clean water and air, allowing polluters to be charged for damage such as adding carbon dioxide to the atmosphere—had emerged in the 1990s (see chapter 3).

Institutions of global governance, in principle at least, made governance of ecological or "ecosystem" services (as they came to be called) seem plausible.[2] Human societies depend on natural ecosystems, and an economy cannot flourish without the support of a society, yet the relationship between these is seldom explicit. An economy *inside* society is something that can be imagined through a national perspective, although this has become more problematic as globalization advances. But once natural ecosystems are invoked, we move beyond the national and take a planetary perspective or a "view from outer space." This view argued for the necessity of ensuring a constant flow of ecological services if the economy was to be able to continue to deliver well-being to society, and this required the Earth system to operate within certain bounds.

The "Planetary Boundaries" paper, as it was soon called, had a spectacular career both as a very widely cited article and as a policy tool, and it helped Johan Rockström win the epithet "Swede of the Year" in 2011.[3] The iconic pie chart (fig. 1) has been republished in discussions of global change to explain the idea that there are boundaries—not necessarily to the physical resources of the Earth on which certain human activities depend, but to how much human impact Earth systems can absorb until their functionality is seriously hampered. This represented a shift away from older perspectives on pollution and limits, which had focused on either localized damage or scarcity. Now the focus became the absorptive capacity, or resilience, in the face of the disturbance of large-scale systems.[4] The sustainability and, possibly, the existence of human societies as we know them was "nested" within the operation of an Earth system. The nine domains in which boundaries were assessed included themes of long-standing concern such as climate change and ozone depletion but also global issues that had received little policy attention, such as ocean acidification and biochemical flows of nitrogen and phosphorous.

The article speaks to the success of Earth system science, the particular expertise for integrating biophysical systems across planetary scales. Most of the twenty-nine authors of the paper were Earth system scientists or ecologists (for example, Rockström himself); both the empirical findings in the article and the underpinning theoretical assumptions can be derived from a framework of Earth system studies. The concept of an Earth system had become an increasingly important strand of environmental science in the period from the 1970s onward. Just like "the environment" in 1948, the idea of an Earth system represented an integrative move, one that became fully visible during the 1980s. In 1986 the concept was launched in a

Seeking a Safe Future   153

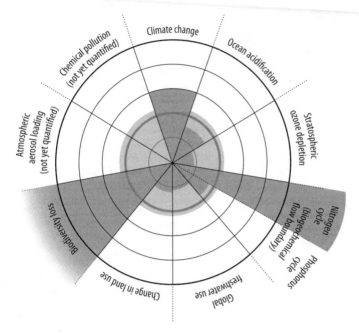

Figure 1. The Planetary Boundaries Diagram summarizes the idea that the Earth system has inbuilt "guard rails," or boundaries, which must not be transgressed lest the Earth system be moved out of its Holocene state of stability and no longer remain safe for humanity. The inner lighter shading shows the safe operating space for nine planetary systems, while the darker shaded areas show estimated positions for each variable. As the diagram suggests, for some of the Earth system indicators the boundaries have already been transgressed. J. Rockström, W. Steffen, K. Noone, et al., "Planetary Boundaries: Exploring the Safe Operating Space For Humanity," *Nature* 461 (2009): 472–75. Original graphics and design by Björn Nykvist.

NASA document and around the same time entered the planning of the International Geosphere-Biosphere Programme (IGBP) that was launched in 1987, hosted by the International Council for Science (ICSU) and, like Rockström's center, based in Stockholm (see chapter 5). ICSU, a union of national scientific associations working internationally, played a role in many integrative efforts, including

UNESCO, the International Geophysical Year of 1957–58, and the International Biological Programme for the decade from 1964. The approach of the IGBP, notably integrating the biosphere with the geosphere, each with its own range of sciences, soon became institutionalized as a common frame of thought. Over a period of two decades, three other "global programs" also formed: the World Climate Research Programme (started in 1980 by the WMO and ICSU in collaboration; see chapter 5), which also informed the Intergovernmental Panel on Climate Change (IPCC). The Earth system also formed the intellectual backbone of the Amsterdam Declaration in 2001 organized by the IGBP, with the central idea that the Earth is "a single, self-regulating system, comprised of physical, chemical, biological and human components." It was followed by a London Declaration, adopted during the conference A Planet under Pressure in 2012, which in turn was an essential step toward the formation of the new global architecture for a science research platform for global sustainability, Future Earth, which started its activities in 2015. One of its global hubs was, once again, placed in Stockholm, the other four in Colorado, Montreal, Paris, and Tokyo. Notably, in line with science funding but out of line with the distribution of the world's population, the hubs were all located in prosperous regions of the global North. Future Earth is a federation of projects, knowledge-action networks, and other initiatives related to Global Environmental Change, with activities in places well beyond its main hubs.[5]

Earth System Science Thinking

One might think that the environment is conspicuous in its absence from this blizzard of world-bestriding acronyms and evocation of spheres and Earths. But in many ways these developments were the culmination of processes that were accelerating in the environmental age after 1948 and that have, as we have shown, longer histories behind them. These global programs had prediction at their core and, equally, the capacity to transform information into numbers that could be integrated into global models. They required the aggregation of an extraordinary amount of expertise, contributory and not least interactional, to generate consensus and shared approaches among thousands of scientists. At such levels, the twenty-nine authors of "Planetary Boundaries" might be seen as an almost immodestly small team.

And just as the environment had, only even more so, the Earth system evoked a planetary scale, but with a different kind of scalability: unlike the environment both local and global, the Earth system always stood as one planetary unit, disarticulated into "components" or "dimensions." The idea of the Earth system became quickly embedded in institutions that had their origins in the postwar decades of integrating the various dimensions of the environment. Perhaps this was precisely because it was a fulfillment of the promise held out by the techniques and expertise developed over that time. Still, to claim that the environment in its global guise was not just an environment but also a system with complex, vulnerably interrelated parts was to take the pattern of thought a major step forward. A system is, by definition, calculable as an entity. And now the entire planet, including its atmospheric cover and the life in it, on it, and over it, was a system that could be calculated and hence thought of as within reach for human organization, planning, and care.

Older notions of this Earth systems thinking, some already familiar to the reader of this book, are sometimes reiterated when the genealogy of modern Earth system science is sketched. Thinkers like Carl Linnaeus, Georges-Louis Leclerc (the Comte du Buffon), Alexander von Humboldt, and Justus Liebig (among others) had declared the interconnectedness of the phenomena they studied in the world. A little later in 1864, George Perkins Marsh depicted "man" as a comprehensive agent of change in nature and landscapes and frequently used the term *Earth*; he became the patron saint of the Princeton conference of 1955. In the 1920s we find Vladimir Vernadsky, the Russian geochemist, and his biosphere concept. Vernadsky would be a major inspiration to Soviet modelers of the environment in the 1970s and 1980s.[6] In the 1920s we also find the highly influential figure of Alfred Lotka, who perceived industrial society and the world's ecology as one whole linked by energy exchange and an in-

dustrial age as a new "cosmic epoch," where "man has initiated transformations literally comparable in magnitude with cosmic processes."[7]
But these individuals' ideas of the Earth as a place of "economy" (in
the older sense of a kind of integrated household), of circulation and
interconnection, were not enough to establish momentum for a new
discipline in their own time.

The roots of Earth system science as we now know it can be seen
as a meeting between *cybernetics*—whose earliest major tract was
MIT mathematician Norbert Wiener's book by that title from 1948—
and the wider geological and geophysical sciences as they unfolded
in the context of the Cold War. The International Geophysical Year
1957–58 and the planning for that event in previous years was an
essential breeding ground, especially when it came to realizing the
productive benefits of large-scale infrastructures for monitoring and
data collection and computer-aided processing. Much of this work
had strategic and military justification and, not least, funding. And it
was the IGY that inspired the notion in Solly Zuckerman's mind that
there should be something called "the environmental sciences."[8]

As we have seen in chapter 4, cybernetics, the study of control
systems, had already made a deep imprint on ecology. The subdiscipline of "systems ecology" was a major factor in the formation of Earth
system science, while other spin-offs such as computerized global
modeling became essential tools in thinking about climate and economic development. Yet Earth system thinking took a long time to
coalesce into a coherent frame of thought beyond those evocations
of connectivity that were themselves centuries old. How could such
an insight be meaningfully turned into systematic and perhaps predictive activity?

One of the early attempts to formulate the idea of the Earth as a
self-regulating system—physically, chemically, and biologically, and
subsuming humans into these categories—came from the British
physicist James Lovelock. Lovelock had worked as a missile expert
during the Cold War and was thus deeply familiar with cybernetic
feedback thinking in precisely the style laid down by Wiener, who
had clarified his ideas while working on antiaircraft fire in the 1940s,

or Jay Forrester, the patron of the World3 model, whose wartime work also focused on missiles. Lovelock subsequently worked for NASA, researching the possibilities for life and habitation on Mars, encountering the challenge of thinking about a whole planet in a connected way.

In the early 1970s, with the assistance of biologist Lynn Margulis (who published *The Symbiotic Earth*), Lovelock published several articles arguing for a view of the Earth as a self-correcting, homeo-static system binding organic life and the inorganic elements; he called the Earth Gaia, a name derived from the Greek Earth goddess. Similar thoughts of benevolent self-regulation had previously been proposed for society by thinkers such as Italian Vilfredo Pareto (a socialist) and Talcott Parsons (an admirer of market liberalism), both in the interwar years, when the concept of homeostasis had a period of popularity and was then taken up by the systems ecologists. When Lovelock scaled them to the level of the entire Earth they became controversial, although Lovelock's personal manner probably played a role in the reception of his work. He was perhaps not an institution builder, nor much of a diplomat. The Gaia idea (or metaphor)—the name actually suggested by his village neighbor, Nobel Prize–winning author William Golding—became as much a point of contestation as one of crystallization; it resonated through the media and to a wider public but lacked the backing of "interactional experts" among lead-ing scientists who could bring institutional support and funding. Other key scientists rarely cited Lovelock, who was much criticized by some. His repeated assertion that the chief problem of Gaia was its human population, which he felt should be radically reduced, was politically unfashionable during the 1970s when arguments moved against neo-Malthusianism. Indeed his association of general argu-ments about population and planet and the need for immigration control in Britain was reminiscent of the views that soured the rep-utation of Garret Hardin around the same time.[9]

Earth system science and thinking thus coexisted with, and was indeed part of, the formation of the environment but was only real-ized as an institutionalized body of expertise and calculation at a certain moment of institutional and technical development, paralleled by new institutions for global governance (both by sovereign states and NGOs, as seen in the previous chapter). In this it had similarities with the issue of climate change, with which it shared dependence on computing capacity and roots in planetary scale geophysics that had links to the military and security establishments. The field-based, "green" dimensions of environmental expertise were very different. They were based on small-scale agents and were more easily mobilized, serving as the vanguard forces of the concept's diffusion. Earth system science arrived in the shape of environment's Big Science, which also meant access to new infrastructures and funding sources and new relations to power.

A Governable Earth?

Earth systems thinkers embodied many of the framing themes that we have emphasized in the previous chapters. To begin with, they quickly formed a particular meta-specialization, a kind of transboundary, collaborative expertise. They had a deep trust in numbers and worked to produce and promote numbers about the Earth perhaps more intensely than any other strand of environmental expertise. They labored constantly to connect locally gathered data with geophysical and geochemical models of all the Earth's big systems, or environments, the oceans, the atmosphere, and the Earth's terrestrial parts. In addition, in the 1970s they made a strong case for including the cryosphere (ice and snow) and for extending the analysis beyond the atmosphere to even more distant spheres of interplanetary space, which became integrated into a globally scaled system of monitoring. This system incorporated large-scale monitoring equipment such as rockets and satellites (Landsat was launched in 1972), submarines and buoys in the oceans, and ablatographs and other devices on glaciers that measured their retreats and (more rarely) their advances. As the ambition and programming power of Earth systems scientists grew, so did the scale of their equipment and the complexity of their models.

This infrastructure, partial as it still was for many years, lent credence to the value of their work and, eventually, their results. What was less prominent in such work—or in fact pretty much ignored—was the complexity of humans and societies. It was an immensely complicating factor and a dangerous one to include in the equations. This may explain why this work kept its distance from environmental*ism* or anything else that could somehow undermine the credibility of the emerging and provocatively far-reaching systemic *Weltanschauung*.

The explicit emergence of Earth system science in the 1980s may partly be explained by a shift in effort as the Cold War funding of geophysical and space research declined. The withdrawal of the United Kingdom and the United States from UNESCO under the Thatcher and Reagan administrations also possibly played a role. Alongside these more opportunistic explanations, some in the global change science community wanted the ICSU and other major players to organize a longer-term effort to move beyond dealing with the "crisis of the month" and instead "establish a comprehensive scientific framework" for the kinds of crises that global change would inevitably bring about. This perhaps reflected frustrations with the progress of environmental policy similar to those that had led to the Brundtland Commission earlier in the same decade (see chapter 6). In other words, the creation of Earth system science was also a political maneuver; this was a science that aimed to get things done, and this required some match between institutional capacity and intellectual ambition: the chiming together of contributory and interactional expertise.[10]

While not denying the importance of more immediate, short-term explanations that have also been characteristic of the "auto-historiographies"[11] of members—individual and institutional—of the Earth system science community, the phenomenon can be placed within a wider historical framework provided by the concept of environment. The UN Stockholm conference of 1972 had demonstrated that the concept had reached, as we have seen, the highest pinnacles of global diplomacy and prestige. Still, surprisingly little was done with it in real terms, especially outside rather traditional practices that were beefed up by national legislation and enforcement, such as pollution control and conservation. Most things continued just as they did before. The upward trend of Charles Keeling's Mauna Loa curve was unbroken. So were most other indicators, whether for ocean acidification, biodiversity loss, nitrogen dissolution, or others. Indeed, they were almost invariably advancing exponentially. The 2005 Dahlem conference (named after the wealthy part of Berlin where the conference villa was located) issued a broad survey of these key, and apparently dismal, environmental indicators, leading to the idea of the "Great Acceleration": that there had been a sudden take-off of human impact on the environment since 1945.[12]

What was to be done? The Brundtland report that we examined in chapter 6 was one attempt that strived to integrate its discontents with approaches to management and governance to inject new life and urgency into conceptualizing the environment. Much of the thinking was taken from a familiar economics toolbox, circling around the creation of incentives for more sound behavior. By the 1990s, this would lead to suggestions for creating quasi-markets for natural resources and commodifying finite environmental qualities such as fresh air as so-called ecosystem services. Aspects of Earth system thinking mirrored this governance in the environment. In a short paper, the French philosopher Michel Foucault, who died in 1984, launched the concept of "governmentality," which has been widely adopted in the twenty-first century.[13] As always his thinking took a long-term perspective, analyzing how modern societies include ever more complex and natural dimensions of the world and bring them under some sort of managerial and decision-making regime. Now the time had come for the Earth itself, the entire planet. Foucault's argument helps us understand this shift.[14] If the concept "the environment" was about the discovery that humans were changing nature in profound, lasting, and disturbing ways—a radical idea in 1948 even if having many precursors—Earth system science was about making the environment on the largest scale *governable*, that is, the object of policy, measures, legislation, and incentive schemes in order to achieve its long-term sustainability.

In order to make things governable they must be visualized and organized, made legible and envisaged with sufficient information to make such steering possible (theoretically, at least). This was achieved through texts, clocks, numbers, tabulations, maps, and similar devices and, ultimately, technologies. In the case of *biopolitics*, to invoke another term that Foucault invented, Earth system science produced similar technologies, primarily computer models and diagrams of how the different properties of the Earth functioned together. The very first "code" of Earth system science for the IGBP was a diagram presenting the interconnections between systems—chemical, physical, biological—with a small box for humans at the far end of the scheme. This was named the (Frank) "Bretherton diagram" after the chairperson of the IGBP planning committee that presented it (fig. 2).

So-called integrated Earth systems modeling became a chief activity of the global programs, developing rapidly from what seemed, in retrospect, rather primitive beginnings in efforts such as the *Limits to Growth* report of 1972 (see chapter 3). These models produced, in the coming years, the kind of perspective that the famous and alarming "hockey stick" curves represented—a quantified, modeled version of Vogt's problem catalogue of 1948. But, more than that, it was a vision that appeared governable, precisely because of its use of expertise, its sovereign trust in numbers, its obsession with the future and its projection, and its confident assumption in scaling: that is, what is local is also global and the global thus also tells the story of the local. It projected the possibility of *geopower*, a word coined by Jean-Baptiste Fressoz and Christoph Bonneuil in reference to Michel Foucault's notion of "biopower" as the means by which various technologies were employed to exercise control over populations via health and penal policy since the Enlightenment.[15] Geopower, then, represents control of the planet, a power that requires its own set of technologies and instruments—and certainly raises new problems of *governance*, since the units and reach of different sets of political power in no way scaled as smoothly as the Earth systems models did from the atom to the atmosphere. Governing societies seemed to always fall somewhere in between.

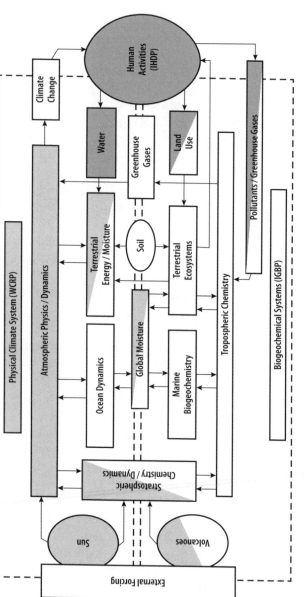

Figure 2. Version of the "Bretherton Diagram" produced by the National Center for Atmospheric Research. The original diagram was developed by Frank Bretherton as a model of the functioning of the Earth system in the 1980s and published in a report for NASA. Harold A. Mooney, Anantha Duraiappah, and Anne Larigauderie, "Evolution of Natural and Social Science Interactions in Global Change Research Programs," *Proceedings of the National Academy of Sciences* 110 (Supp 1, 2013): 3665–72.

It is not as if knowledge stood still for all this time and change was simply a matter of integration and presentation. The contributory disciplines of the environmental sciences toiled away at their research, and the collators and calculators who constructed global models stimulated work on new foci and brought significant connections to light. The nine dimensions of the "Planetary Boundaries" paper represented a reworking of the problem catalogue and a re-envisioning for a digital and globalized age.

A Safe Operating Humanity for Humanity

It took many years until Earth system science could begin fully to translate the modeling into more concrete suggestions about how to think in terms of policy tools. Indeed, the notion of geopower remains largely a fantasy, whatever kinds of policy instruments one might imagine, from radical reforms of energy markets to geoengineering. There was, nevertheless, a relatively rapid recognition (or fear) of the political potential of this work. Strong critiques emerged of envisaging the global system, and potentially global politics, from a "point of nowhere" and with little care for human and social realities, which were only crudely modeled in this system, if included at all. These were echoes of divides between rich and poor nations and arguments about reductionism that accompanied both the Stockholm conference of 1972 and the publication of the *Limits to Growth* report in the same year.

A critique from the right of the political spectrum, especially in the United States, Canada, and Australia, rejected evidence of climate change and global change more generally as a plot to impose forms of technocratic government, perhaps on a world level, reining in liberty and enterprise. On the left, the global models were perceived by some as the precursor to imposing forms of technocratic government, perhaps on a world level, to ensure the continuance of a capitalist economy rooted in managerialism and inequality. The environment, starting with all the apparent innocence of fact, had now moved to the sphere of values, for which Earth system science, with its models and diagrams based on thick layers of numbers, did not seem very well equipped after all.

Of course, as we have seen throughout this book, the concept of the environment was *always* about humans, in the different forms that it took, and the struggle over what values should govern societies is not new, with differences often being deep and fraught. What kinds of landscapes should be preferred? Who is to sacrifice or pay for environmental costs? Isn't there also environmental justice alongside the technologies of environmental managerialism? The very same questions could easily be scaled to the global level, and with time they were. In fact, isn't the wealth in certain parts of the world resting precisely on the environmental "slow violence," to use Rob Nixon's expression from 2011, that is exerted toward the "wretched of the earth," to use Frantz Fanon's phrase from 1961?[16]

The aggregation of environmental problems into global dynamics as presented in Earth system science posed new questions about the force of collective humanity—or whether there was even a human collectivity at all. This was already recognized when global programs, attempting to include human factors as variables like biophysical factors, discovered serious opposition from humanists and others outside circles of modeling expertise. Earth systems theorists used terms like *human* or *societal dimensions*, but these were not expressions that reflected the ways scholars of human behavior or society described their subject matter. They seemed to imply that humans and their interactions could be reduced to a subsystem. The International Human Dimensions Program (IHDP), sponsored by, among others, the International Social Science Council (ISSC), had provided an alternative kind of knowledge production in many of its activities. IGBP's subprogram, Analysis, Integration and Modelling of the Earth System (AIMES) hosted the Integrated History and Future of People on Earth (IHOPE) project, yet another, and perhaps not entirely welcoming, acronym.[17] IHOPE has since 2003 worked to include past, more than present, human societies in our understanding of global change, often on timescales beyond those typically dealt with by historians, and sometimes developing metrics that supplemented or tweaked established approaches in economics: measuring human development, natural capital, ecological services, and the like. These have run parallel to an emerging interest in "Big History," led by scholars such as David Christian and Ian Morris, who work on a scale well beyond usual (and everyday) assumptions about agency, politics, and experience.[18] Such attempts at integration and synthesis of approaches to long-term change in different disciplines have often meant an extension of the discourse and practice of the natural sciences into new areas—in this case, the writing of history and the social organization of time.[19] They fit together timescales hitherto separated by disciplinary traditions and incommensurable thought patterns into a single narrative of "planetary" time—what we might call the "Great Synchronization."

Often advances in thinking and a reorganization of knowledge and outlook bring with them new concepts that seek to better capture the essence of the new ideas. Earth systems thinking is no exception. Nobel Prize–winning atmospheric chemist Paul Crutzen silenced an IGBP workshop at Cuernavaca, Mexico, in 2000 when he interrupted discussions about human-induced changes on multiple aspects of Earth systems to say quietly, "we are not in the Holocene anymore." We are in the . . . *Anthropocene.*[20] The Holocene, as we have seen, was the geological epoch of the past 11,700 years in which agricultural societies emerged and flourished. The Anthropocene, the epoch of humans, is a new age where our actions are at least as significant as any other force in driving global change across all those dimensions that Earth system science has spent the past decades describing and measuring.

From this impromptu intervention, Crutzen developed the idea into an article with fellow scientist Eugene F. Stoermer in 2000, and the idea gathered pace in the Earth system science community, particularly after its publication in *Nature* in 2002.[21] The idea has now spread well beyond geology and Earth system science. In times of rapid international and global change, it has become a metaphor for the present (and future) in the humanities, for artists, in the popular media, and in the lexicon of politicians.

The idea of the Anthropocene as a geological era is not just another example of human solipsism, people turning the history of the planet into their own history, or indeed marking a moment in human history as if it had geological significance. The Anthropocene is an argument that human influence—and the traces of that influence—have now irreversibly altered the Earth system and will continue to do so whether we humans exist or not. The anxiety about the human legacy on the planet is not just coming from earth scientists. British nature writer Robert Macfarlane weighed in with his own take on humanity, declaring: "'What will survive of us is love,' wrote Philip Larkin. Wrong. What will survive of us is plastic—and lead-207, the stable isotope at the end of the uranium-235 decay chain."[22]

Nevertheless, the Anthropocene is also an epoch in our own history, and arguments for it represent a moment in the development of human knowledge about the environment. The markers of profound human influence are found in measurements taken in the oceans, in the atmosphere, of changes in biota, of radiation and the detritus of consumption and construction. Neither specialized research disciplines operating in isolation nor even the most accomplished of polymaths could have detected that such changes were more than widespread; they were changing the way biophysical systems work on Earth. Such change is only discoverable via the *aggregated* expertise by which disciplines and observations have been integrated, and interactional experts have constructed models and mapped change. Prophets of the past, such as John Ruskin in England or Eugène Huzar in France, both in the nineteenth century, may have decried and feared the destabilizing influence of humans on their world.[23] But the knowledge of the Anthropocene is different and bears with it a different kind of authority and predictive traction. Perhaps it also holds out the promise of a new understanding of time itself and the roles of humans in it, a major rupture in philosophical perspective—Earth system science in the Anthropocene as a *Weltanschauung*.

Every view is constructed *somewhere*, even if the vision of the globe is all-encompassing. There are hierarchies of collection, calculation, and communication. And if Earth is one "system," and humans a steering element within that, what does it do to the intractable individuality and diversity of humans? Aggregation is certainly not the same as pluralism, and seeing humanity as a species is rather different from seeing it in terms of society. Of course, as we have seen (especially in chapter 1), the emergence of the environment was also accompanied by quite generalized claims about humanity and its impact. But throughout the decades after 1948, human influence on the environment was a factor that could be applied or withdrawn. Now there was no going back. Thus the Anthropocene has seen the development, more forcefully, of two elements of disputation and unease in environmental discourse.

The first element relates to scientific authority and the configuration of relevant disciplines, and hence identification of the relevant information for understanding the Earth system or any part of it. If human action is now irreversibly part of the environmental story, what did it mean for the techniques of prediction, aggregation, and policy that had been developed in previous years? Could "humanity" become part of a climate or Earth system model in any meaningful way? Equally, others felt discomfort at the apparent politicization of science. Was the declaration of the Anthropocene itself not a political gesture, one that would cascade through scientific practice? Predictive models could at best reflect the possible outcomes of certain political choices.

This is perhaps felt most intensely among the groups charged with giving an official imprimatur to Crutzen's proposal, the stratigraphers within geology. In scientific circles, it presented a curious intervention: the suggestion to approve the name for a new geological epoch arose from the perspective of Earth system science, rather than from geology in isolation. This led to heated debates within geology and stratigraphy as to what might count as the evidence in the rocks for the Anthropocene. The chair of the International Commission on Stratigraphy was among those who expressed discomfort about a political tail apparently wagging a geological dog.[24] Within the expert community, some have been uncomfortable with adopting a proposal that seems (to them) to come from outside the sober inductive practices of their trade, or that relates to so recent a timespan. As we write, the debate continues and the idea of the Anthropocene has still to be ratified by the International Commission on Stratigraphy (ICS), the institution responsible for the official International Chronostratigraphic Chart that lists the ages of the Earth.[25] Nevertheless, these debates have certainly raised the profile of stratigraphy, while its traditional profile as the apparently neutral, solid arbiter of ages of the Earth could give weight to the idea of the Anthropocene.

Thus stratigraphy has been drawn into a wider politics of the Anthropocene and its implications and made a partner in an Earth system science. Of course, stratigraphers had always relied on interaction with other disciplines to explain and demarcate their findings, but now a much wider integration—perhaps even with politics and humanities—is at stake. The standard practice in stratigraphy has also been to define a point in the Earth—a "golden spike"—which delineates the division of epochs (between, for example, the Triassic and Jurassic). This has been more widely interpreted as the need to identify a "date" in human history as the starting point of the Anthropocene. Practitioners of different disciplines have come up with answers strangely reminiscent of their major scientific interest: scholars of early agriculture propose the clearances of the Neolithic, atmospheric chemists point to the impact of the carbon-spewing steam engine, and some Earth systems experts prefer the very widespread signal of atmospheric nuclear testing and the postwar moment—around 1948—where so many indicators of environmental impact show an upward tick.[26]

A question about what this concept might imply in terms of authority and power has led to a second and related but rather more widespread tension. The discussion of the Anthropocene has reopened questions of expertise for the future and made space for unexpected new expertise—for example, art—not usually associated with matters scientific. As noted above, some suspect that the idea of humans steering the Earth system is a Trojan horse that would justify experts establishing a wider technocracy and managerialism of the environment (fears that reflect, in turn, wider political currents). To some, the Anthropocene opens the door to the reinforcement of the vested interests of the global organization of governance that threaten restriction of liberty and control of markets. The Anthropocene is but a cover for a politics of redistribution and collective controls. In contrast, others see the idea as a cover for precisely such vested interests, the dynamics of the world being determined by global models that require expert-led, antidemocratic solutions that reinforce current configurations of economic and political power. As environmental humanists Marco Armiero and Massimo De Angelis argue, "the Anthropocene discourse conflates the individual and the society at large. . . . If people live in this mess . . . they should only blame themselves as members of the universal human species or, in the optimistic version, act as a member of the same universal human species to improve the situation."[27] Many humanists feel that by generalizing "human nature" to just the greedy and the short-sighted, the Anthropocene idea lets off the hook those responsible for creating this so-called age of humans.[28]

Both kinds of critiques treat the concept as if it has a singular intrinsic logic, but this perhaps says more about the critics' own fears. The whole history of the environment that we have presented in this book indicates that the history of a concept is indeed closely related to the development of expertise, of institutional power and dominant imaginaries, and political influence. But the history also demonstrates that there is much contingent on how these processes play out. The adoption of an idea does not predetermine what comes after it. Indeed, one might argue that the very range of debate around *Anthropocene* indicates that it cannot operate as a term that simply leads to the closure of debate and the dominance of certain groups; on the contrary, it is already a sign of a rupture, a new space for contestation, creativity, and debate. It could also be read as a broader call for "more-than-expertise," for participatory decision making and international collaboration on a scale never seen before.[29] The contestation of the validity and meaning of the Anthropocene has partly arisen in relation to political fault lines already present, sometimes of long standing. How can one talk of "human" responsibilities or humanity as a driving force when there are such disparities in consumption between north and south, between different classes, among women and men, between the histories of nations (although all of these are also entangled in very complex ways)? Is the "culprit" not, rather, capitalism, imperialism, neoliberalism? In the past, in contrast, those who wrote about damage to the natural world did so from elite circumstances in the West and often blamed the damage on nomadism, climatically determined character, religion, and so forth. All of these arguments imply certain understandings about the drivers of historical change and the causes of inequality (not least certain inequalities themselves as drivers of change).

The fear is, perhaps, that a putatively new politics of the Anthropocene only serves to obscure what these objectors consider to have been important about politics all along. Such arguments cannot be easily dismissed; indeed, what is perhaps most significant for our understanding in this book is that they point to the irremediable differences of opinion that are certainly difficult for an Earth systems model to encompass. At the same time, one might recognize that the foregrounding of human agency and contingency is itself maybe no more than an assumption (certainly a few scholars think this), nay, a political preference that can be subjected to empirical testing as with any other one about the world ("natural" or "social"). One thing we cannot escape in these debates is that the Anthropocene is a child of the environment. It is as much about what it means to be human as about the changing planetary systems. This history suggests that the same was true of the environment when it emerged in the 1940s.

Three Ages of the Environment

"The environment" is one of the great concepts of our time. It has lasted longer than the lives of most world leaders today. The concept guides reflection on ongoing processes that have shaped the concerns, institutions, and experts that shape our politics and lives today. One very simple observation is that *environment* as a word (aside from the endless set of things people have attempted to denote and link to it) has a history nearly two centuries old in English (as outlined in chapter 2), and the concept of the environment has been around for seventy years. Yet within these periods it has also been a chameleon and a changing concept, emerging in many different and diverse contexts. It is associated with various techniques of measurement and conceptualization, of authorized and authoritative speech, of politics and the political (with an uppercase and lowercase p) and constellations of contributory and interactional expertise. In all this time, the word has exercised a protean effect and always been itself shaped, *deployed*, in context.

The focus of this book has been the environment rather than environmentalism or its associated activist social movements. Even with this biophysical focus, it is possible to trace how the concept has been shaped and colored by historical and political contexts within sciences and beyond. Formed in the historical period from the 1920s, it gathered pace sharply in the late 1940s, and more recently its strong association with systems approaches has drawn it into "planetary thinking." It is still used and useful well into the twenty-first century for very local phenomena such as workplaces, homes, and urban space.

Yet the collection and processing of quantitative data as a means of rendering the environment legible was fundamentally conceptualized between (roughly) 1948 and 1972, that is, rather earlier than the activist response and rise of environmentalism as a mass political force. In chapters 3 through 5, we have traced the various fields in which it was deployed, gained traction, and had integrating effects, which is also a story of reciprocal shaping, a history of science, technology, and Cold War politics. This postwar period saw many forms of expertise being profoundly shaped by cybernetics, computing technology, experiences in fieldwork, and the rapid expansion of higher education and government funding in the West and beyond, but they also came to be combined in different and at times unsettling ways. By the late 1960s, however, the environment was a domain around which one could organize a conference, demand a policy, appeal to the public, and step forward as a public figure who could speak for an accumulating body of knowledge—while insisting, increasingly, that this was a new kind of knowledge that reached far beyond traditional disciplines and practices. Much of this work was done in academic circles, although with a close connection to older concerns: conservation, pollution control, fears of overpopulation, resource management, or urban planning, matters that (in the view of the new environmental thinking) could hardly be seen in isolation.

From around 1970 we may see the history of the environment evolving in a second (or mature) phase that has accompanied the emergence of ministries/departments of the environment, the United Nations Environment Program (UNEP), a plethora of environmental nongovernmental organizations, and a sometimes bewildering new alphabet of acronyms denoting attempts to integrate scientific and policy communities. Its public recognition was such that it became the focus of new political groups—Green parties that appeared first in Tasmania and New Zealand in 1972 and most prominently in Germany the following year. This was a phase of institutionalization and an embrace of questions of governance and thus an expansion that touched upon more political and social questions, although moralized views of human-nature interactions were central to the pioneers of 1948. The notion of the environment as a planetary or "world" problem, locally ubiquitous, was now also associated with the emergence of a hazardous globe in an interconnected sense, of endlessly iterated action at a distance, alongside economic globalization and questions of what international order would succeed the Cold War. In 1972 we saw a sober and official Stockholm conference, where environment and development were coupled in the new organization UNEP, while protestors and social movements were corralled into fields at the edge of the city.

With the Stockholm conference, we could talk of the emergence of global environmental governance, later with climate as a significant component, which, a half-century later, has grown into a profound and inevitable dimension of world affairs. This was at the same time both a high politics of diplomacy and an increasing mobilization from below. The Rio "Earth" summit of 1992 sought to bring a vast array of actors and activists into the heart of the proceedings—or at least a little closer to them—aspiring to a response to environmental problems that would seamlessly span the local and global. The era of institutionalization, which had lasted over two decades, now faced a new sympathy for pluralization. Ideas of environmental justice, the dilemmas of development, and the distribution of blame became familiar and quite entrenched aspects of environmental debates.

Yet the age of pluralization of the past two-and-a-half decades has not found the inheritance of previous phases of conceptualization and institutionalization easy bedfellows. World summits at Johannesburg (2002) and Rio (2012) were redolent more of disappointment than dynamism. High-level diplomacy on climate change seemed largely to have stalled (for example, the responsibility for historic carbon emissions featured prominently at the Copenhagen summit of 2009 as a reason not to set agreed-upon targets for emissions reductions), while the comparatively new language of Earth systems and global change was distant from grassroots initiatives, campaigns for "transition," and an awareness of economic development that continues to eat up the world's resources at a dizzying rate, despite the tribulations of financial collapse and recession in many regions.

Is the promise of environment politically exhausted? Certainly we can suggest that the use of Anthropocene among the global change "community" of experts is a conscious effort to inject urgency into the policy process, specifically to consider futures bigger than just those affected by climate change, though this remains a strong thread across the field.[30] Equally, the challenges and dissent that have emerged might be seen as one aspect of our present discontent in a time of pluralization, a resistance to an integrative expertise that seems too all-of-one-kind, and an appeal for an inclusion of a much wider set of voices and perspectives. The fact that experts of different kinds (environmental, economic, managerial) look so similar with regard to their techniques of assessment and visualization, their interactions and communications, may give rise to the suspicion that they are not so different in their goals and practices—undoubtedly a simplification but also a reflection of how a culture war has opened up over scientific expertise and its relation to politics that was not really present in the 1950s and 1960s. Perhaps it reflects a meta-expertise of modeling and systems analysis that transcends disciplines but nonetheless limits the possibilities for the future.[31]

It is, as with so many historical judgments, too early to tell, but we might hypothesize that Anthropocene finds itself as a concept in a position similar to the environment in the era between 1948 and 1972. It appears to demand new alignments of old disciplines; some enthusiastic advocates talk of changes in knowledge regimes, some see it as a tool for shaking old orders, others wonder if it merely disguises old vested interests. Who contributes? What forms of calculation and interaction will it create? Will it, in addition to drawing people into debate, generate new experts and presiding synthesizers and seers? Or is it a flash in the pan—destined to fade away? Aspects of the Anthropocene debates, which bring "politics" and "science" into uncomfortable proximity for some or which see Earth systems scientists express a need for more arts and the humanities, might become as commonplace, as utterly ordinary and unquestioned, as the discourse of environment is today. We no longer question that the world is an ecological web under threat from human action, that local and global are connected, that growth poses a host of challenges for humanity, that environment is also a question of security and diplomacy, or that human behavior is somehow misaligned with the capacity of the planet to support it.

All these notions, most in themselves not novel, were bound into what now seems to be a natural and unshakeable set of associations in the quarter century that followed the awakening year of 1948. Yet much has changed, too, and quickly. Chapter 2 compared the Man's Role in Changing the Face of the Earth conference held at Princeton in June 1955 with the Future Environments of North America conference a decade later, to show how much the same aggregated group of experts consolidated their conceptual frame in that time to create a new orthodoxy. Now we can look back over a much longer period at the participants and division of labor at Lake Success in 1948, across 1955 and 1965 (the year of the first major presidential environmental report in the United States), to Stockholm 1972, Rio 1992 and 2012, or Paris 2015 to chart the enormous change in participation and expectation and increased plurality of voices, which all, in their own way, feel "the future in their bones."

The Futures of the Environment

The environment was a work of integrative ingenuity. Still, for some time the mix of what was considered environmental science remained limited even if the scale on which it was conceived was large. The domain of relevant knowledge has grown over time: the environment has become bigger. In the era of the Anthropocene, there is a new, forceful wave of integrative knowledge rolling forth, this time engaging the social sciences and the humanities on a scale hitherto not seen. This is no coincidence. The Anthropocene is proposed as a new geological epoch, but while the International Commission on Stratigraphy (ICS) is deciding the fate of the formal idea, it has escaped as a metaphor for our times, as a space to debate human responsibility as well as planetary futures. Environmental philosophers find it morally problematic to treat humans as a "species," collectively. While it is undoubtedly our species that is shifting the geological rules of the planet, it is not every human.

The conversations around the idea of Anthropocene as metaphor signal a new twenty-first-century shift of emphasis, in our ideas about "the environment"—from the "big thing out there" that humans tweak and maltreat to the "planet as a whole," which we have already irreversibly changed and have to accept *as our own given, but in part a gift from ourselves*— which we still tweak and maltreat. In the first case, or phase, of the environment, focus was on the rate and direction of change. This mobilized a range of knowledge primarily from the natural sciences, the experts on what happens when humans interfere with natural systems, on any scale, from the very small—cells, particles, toxic molecules—to the very large—biodiversity, ocean acidity, greenhouse gases in the atmosphere.

In the second decade of the twenty-first century, interest has shifted from the negative transformation of the environment and the hope of some restraint to what might be a possible, indeed necessary, positive transformation of humanity that might steer the fate of nature. This goes beyond choices, for example, to use or not use some technologies, or whether to tread lightly or wield some big geo-engineering. In Herbert Spencer's time, 150 years ago, *environment* was that external thing that could shape our interior life (see chapter 2). Now, only societies that have themselves changed can manage the humanity-planetary (or nature-culture) relationship. Somehow society and environment have to be brought into alignment as humanity becomes a geological force and there is no longer a distinction between human history and natural history.[32] These circumstances invoke the need for the human sciences, the experts not of environment but of people, cultures, and societies. It is, at the same time, a new kind of humanities, mobilizing not only traditional skills but also art and performance,[33] transforming in the face of crisis, just as the natural sciences started to change, reorganize, and integrate in new constellations as the long and demanding "problem catalogue" of global environmental issues grew more extensive from 1948 onward.

What kind of knowledge and humanities might this mean? This is by no means clear and might take varied forms. We cannot engage in a new history and prophesying for the humanities here, already at the end of a long and complex story. But we can identify the key new elements that the humanities introduce. In brief, the tasks for the environmental humanities are to situate the human in geological terms and to situate the nonhuman (or "more-than-human") in ethical terms.[34] How this role is understood can be divided into three differing approaches.

Firstly, for some the role of the humanities goes no further than being effective communicators, acting as persuaders for the necessary fixes that science and technology will provide ("as they always have before," a techno-optimist might add). This approach for the most part holds to the "deficit model" of science: that is, that people behave in ways that scientists think is unwise (as if scientists all agree on such things) simply because they do not know enough about science. Educate, explain, cajole, and people will come around. Such thinking implies that there is no real debate to be had about *values*, because all problems can be resolved by *more information*. Evidence for the success of this approach by itself is thin, and we have spent much of this book explaining how the environment is a matter of *imagination* as well as information, equally among scientists as among nonscientists.

A second option is for the humanities to get in on the game that already exists: to integrate with environmental expertise in the form that has emerged in the past decades. Many of the techniques and habits we described as part of contributory and interactional expertise are not exclusive to the natural sciences, and many, such as trust in numbers or scalability, are well established in the social sciences that have a parallel history. As we write, the Nobel Prize–winning economist Amartya Sen of Harvard University is the leading figure in an IPCC-emulating worldwide team of some three hundred social scientists and humanists engaged in the International Panel on Social Progress. Here we see a very conscious effort to produce "aggregated expertise" for political effect.[35] Other approaches mimic the strategies of Earth system science thinking and modeling more directly. If the idea of ecological or ecosystem services was a means of integrating the value of nature into the discourse of economics in a way that allowed these domains to be modeled together, why not also have *cultural ecosystem services*, a measure of the way nature allows value and meaning to be generated in society? Such integrations will certainly have their consequences, privileging experts and metrics that have command of particular techniques and forms of interaction. Aggregation of expertise in the past has tended to lead toward new meta-specializations and modes of institutional power, as we have extensively described in this book. It would be surprising if such patterns did not repeat themselves, and one must ask again, whatever their benefits, who is privileged by such means?

The past two decades have seen the emergence of a third strand, an explicitly environmental humanities. Pioneering work was done in Australia, where anthropologists, historians, philosophers, and ecocritical literary scholars, among others, came together to form the field in the late 1990s and the early 2000s. They sought to orient their work explicitly toward the ecological challenges of the age, not simply as a novel theme (which was not after all that novel, certainly not in the guise of "nature") but in a way that called into question the established preoccupations of their fields of study as narrow, even myopic and inadequate.[36] Climate change has undoubtedly been a major factor in bringing these new efforts about.[37] It represents a long-delayed institutionalization of environmental concerns within the humanities, although there had certainly been humanist scholars present in the environmental gatherings of the past, such as in Princeton in 1955, who harked back to George Perkins Marsh (1800–82), himself a literary and linguistic scholar in a more polymathic age.[38] There were also a few early adopters in Europe in Germany, Sweden, Italy, and the United Kingdom and a rapidly expanding field in the United States. One characteristic of many of these initiatives is that they are experimental in form and method, with a certain aversion to traditional "disciplines." Far from being introverted, there has often been a conscious effort to reach toward the environmental sciences. Leading science journals have begun to carry articles carving out a new role for the humanities, such as a seminal article by the former head of the Tyndall Centre for Climate Change Research in Britain, geographer and climate scientist Mike Hulme, whose "Meet the Humanities" appeared in the very first issue of the journal *Nature Climate Change* in 2011.[39] These efforts still remain a very small although growing proportion of work in the humanities. Perhaps thinking that is significant to understanding the environment no longer needs to be explicitly "environmental."

Common to the three strands is the idea of framing a problem or concept as a narrative or story, breathing life into complex or abstract ideas such as climate change, biodiversity, or environmental justice. This is not exclusive to the environmental humanities, and we might say it is an approach that is simply human. But taking it seriously epistemologically is fresh, even radical. The environmental humanities are inclusive. There is no single "right way" to tell stories, but the best stories will ring true for many audiences.[40] The environmental humanities, rather, seek out a diversity of ways to speak on a human scale about our times of rapid environmental change to audiences from within and beyond the academy. They use the skills and tools of a range of disciplines—history, anthropology, literature, geography, philosophy—and then they must *throw away the disciplinary scaffolding* to reach beyond ivory tower readers and listeners. Often artists, musicians, performers, and educators are drawn to the work. The humanities are already taking at least three forms in the face of these challenges. They are not mutually exclusive or necessarily contradictory. They also foreground the issue of *who* is telling the story, *from where* and *about whom?* Can one be a global storyteller, or must we share an ocean of stories?

We might wonder, however, if all these strands are leaping onto yesterday's bandwagon, attaching themselves to a concept whose emergent moment is already decades old. Yet that environmental expertise and the tools it has adopted such as modeling and prediction, instrumentalized by the digital revolution, are found across policy advice both inside and outside government, in economics, environmental science, planning, and disaster management alike. The environment is old enough to have a history, yet still very current. But its manifestation in the policy arena and science has too often become a way to visualize, rather than imagine, the future, as if the work of imagination has now been done. Anthropologist Arjun Appadurai describes the future as conceived in modeling scenarios and managerial theory as "technical or neutral space," lacking recognition that the future is also "shot through with affect and with sensations," and it is these latter factors that stimulate hope and motivate action.[41] He persuasively argues that "the capacity to aspire is unequally distributed" in a world where the future-making is limited to technical expertise. The imagination still remains "a collective practice that plays a vital role in the production of locality" (including the sense of belonging to a global community).[42]

Who is then included in analyses of costs and benefits, and how are such things measured? This is a world of "slow violence" dealt out by environmental disasters brought on by rapid development and growth without checks, particularly in poor places under pressure from rich partners.[43] What risk, uncertainty, wealth, and poverty mean are refracted in a bewildering hall of mirrors, reflecting one world, perhaps, but not one view or experience. If the idea of the Anthropocene is an ultimate "global narrative" beginning with a concern for the changing planetary environment, it soon poses the question of whether humanity really shares one story of its responsibility to the Earth and fellow life-forms. And it poses to us, in different ways, the question of what we collectively are or might be worth. It is in the human interest to save the planet from the extremes of anthropogenic change, but it is not obviously in the planet's interests to save humanity. The Anthropocene is also the age of works such as *The World without Us*—an unsentimental probing into a future where the prehuman past returns in a posthuman future that also sheds some light on how we might imagine the place of humans on this planet.[44]

The environment has a history. That, at least, should be clear by now. But does the environment have a future? The question of predicting its future and managing the environment in the face of future uncertainties have been the focus of science, policy making, and geopolitics since 1948. Facing the future has been the style of Western thinking, increasingly so in the second half of the twentieth century and into the twenty-first. Will the concept of the environment travel into that future? The environment may not end, but it would be surprising if it did not look as different in 2048 and certainly 2100 as it does looking back to 1948 and even more to 1918. It is evolving, as is the expertise for its management. Perhaps even the idea of expertise is shifting. As concerns for justice and local participation in global decision making demand new ways of understanding, alternative possibilities come into play. Some might stop trying to "game the future" and do as many traditional groups do—face their ancestors.[45] If the future is behind us, we face our past. In "Let Them Drown," her provocative 2016 Edward Said lecture, Naomi Klein suggested that the future might depend on alternative, non-Western models. In a warming world, people need to be more than good citizens; they should also be "good ancestors" and take account of the world in seven generations' time.[46] The history of the environment that we have traced here has certainly been dominated (although not wholly) by the norms of Western science and politics of the past two centuries. Conceptualizing the "human" in terms of the immediate present (and in Western terms) normalizes the historically exceptional few years during which the environment has flourished and created ways of seeing our planet unthought of before now.

The environment has been narrated and renarrated, reinventing itself over and over again. Storylines of decline[47] have predominated since the middle of the twentieth century, the age of the environment. Reflecting on the Anthropocene narrative helps to remind us that there were other sorts of "environment" before "the environment." "Nature," for example, has left its stratigraphic traces in our language and understanding. This was an environment of minds, of individuals, of physiological bodies and organs, of institutions, of species. This was the raw material from which an integrated study and comprehension of "planetary" issues was forged. This planetary environment, in turn, was opposed to humanity. Yet life on Earth, including its people, remains stubbornly local, with an environment unique to each.

We do not argue for a return to the thinking of the late nineteenth century, which had its own forms of reductionism, silencing, and worse. But this story of scale, evidence, and trust is one that remains plural and ongoing. There is *the* environment, but there are environments, too. Decades of environmental science and thought have brought us great benefits. Thinking with the environment has changed the world. And future generations might talk about changing it yet.

Notes

Chapter 1. Road to Survival

1. Rachel Carson, *Silent Spring* (London: Penguin, 1963), 23–25, passim.

2. P. Brooks, *The House of Life: Rachel Carson at Work* (London: George Allen and Unwin, 1973), 239, 263.

3. Thanks to David Moon for this observation.

4. See also the discussion by Mary Douglas in "A Credible Biosphere," in *Risk and Blame: Essays in Cultural Theory* (New York: Routledge, 1992), 255–70.

5. Carson, *Silent Spring*. *The Web of Life* was also the major high school biology textbook in the United States in the 1960s and was borrowed and adapted in other places such as Australia. Libby Robin, "Radical Ecology and Conservation Science: An Australian Perspective," *Environment and History* 4, no. 2 (June 1998): 191–208. The phrase itself dates to the eighteenth century.

6. Carson, *Silent Spring*, 69, 79, 168.

7. Ibid., 22.

8. William Vogt, *Road to Survival* (New York: William Sloane Associates, 1948).

9. Ibid., 283, 285.

10. Ibid., 287, 288.

11. Ibid., 14–15.

12. Aldo Leopold, "The Land Ethic," in *A Sand County Almanac and Sketches Here and There*, 201–26 (Oxford: Oxford University Press, 1949), 202.

13. Vogt, *Road to Survival*, 271.

14. Fairfield Osborn, *Our Plundered Planet* (Boston: Little, Brown, 1948), vii.

15. Ibid., 29; italics in original.

16. Paul Warde and Sverker Sörlin, "Expertise for the Future: The Emergence of Environmental Prediction c. 1920–1970," in *The Struggle for the Long-Term in Transnational Science and Politics: Forging the Future*, ed. Jenny Andersson and Eglė Rindzevičiūtė, 38–62 (London: Routledge, 2015); Libby

Robin, " 'The Environment' and Its Evolution as an Integrative Tool," blog for the International Social Science Council (ISSC), under the auspices of the Futures Past collective, Institute for Advanced Sustainability Studies, Potsdam, Germany, 2017.

17. Theodore Porter, *Trust in Numbers: The Pursuit of Objectivity in Science* (Princeton, NJ: Princeton University Press, 1995).

18. David C. Coleman, *Big Ecology: The Emergence of Ecosystem Science* (Berkeley: University of California Press, 2010). On Big Science, see Derek J. de Solla Price, *Little Science Big Science* (1963; repr., New York: Columbia University Press, 1986).

19. Robert Gottlieb, *Forcing the Spring* (Washington, DC: Island Press, 1993), 172; Lester Machta, Robert J. List, and L. F. Hubert, "Worldwide Travel of Atomic Debris," *Science* 124 (1956): 474–77.

20. Lynton K. Caldwell, "Environment: A New Focus for Public Policy?," *Public Administration Review* 23, no. 3 (1963): 132–39.

21. John Sheail, *Nature Conservation in Britain: The Formative Years* (London: Stationery Office, 1998), 160–66; John Sheail, *The Natural Environment Research Council—A History* (Swindon: Natural Environment Research Council, 1992).

22. Discussions about what should be done "about the environment" increased throughout the decade, culminating in institutional developments circa 1970 in many Western countries. For an Australian example, see Libby Robin, with Max Day, "Changing Ideas about the Environment in Australia: Learning from Stockholm," *Historical Records of Australian Science* 28, no. 1 (2017): 37–49.

23. Lynton K. Caldwell, *Environment: A Challenge for Modern Society* (New York: Natural History Press, 1970), 244.

24. Although in fact he did, but only in his early years. Sverker Sörlin, *Carl-Gustaf Rossby, 1898–1957* (Stockholm: Royal Swedish Academy of Engineering Sciences, 2015), 15.

25. Paul A. Edwards, *A Vast Machine: Computer Models, Climate Data, and the Politics of Global Warming* (Cambridge, MA: MIT Press, 2010), 158.

Chapter 2. Expertise for the Future

1. William L. Thomas, *Man's Role in Changing the Face of the Earth* (Chicago: University of Chicago Press, 1956).

2. George Perkins Marsh, *Man and Nature; or, Physical Geography as Modified by Human Action* (London: Low, Son and Marston, 1864); Justus Liebig, *Chemistry in Its Application to Agriculture and Physiology* (London: Taylor and Walton, 1842); Paul Warde, "The Invention of Sustainability," *Modern Intellectual History* 8 (2011): 153–70; Stephen Stoll, *Larding the Lean Earth: Soil*

and Society in Nineteenth-Century America (New York: Hill and Wang, 2002); Benjamin R. Cohen, *Notes from the Ground: Science, Soil and Society in the American Countryside* (New Haven, CT: Yale University Press, 2009).

3. F. R. Leavis and Denys Thompson, *Culture and Environment: The Training of Critical Awareness* (London: Chatto and Windus, 1933); Isaiah Berlin, *Karl Marx: His Life and Environment* (London: Butterworth, 1939).

4. Herbert Spencer, *Social Statics; or, The Conditions Essential to Human Happiness* (1851; repr., New York: D. Appleton, 1883), 80.

5. Ibid., 7.

6. See Herbert Spencer, "The Factors of Organic Evolution," originally in the April and June editions of *Nineteenth Century* (1886), and published as *The Factors of Organic Revolution* (New York: Appleton, 1887).

7. Herbert Spencer, "The Ultimate Laws of Physiology," *National Review* (October 1857).

8. Herbert Spencer, *The Principles of Psychology* (London: Longman, Brown, Green and Longmans, 1855), 194. See also Trevor Pearce, "From 'Circumstances' to 'Environment': Herbert Spencer and the Origins of the Idea of Organism-Environment Interaction," *Studies in History and Philosophy of Biological and Biomedical Sciences* 41, no. 3 (2010): 241–52.

9. Herbert Spencer, *The Principles of Sociology* (London: William and Norgate, 1876), 6. See also Herbert Spencer, *On Social Evolution: Selected Writings*, ed. J. D. Y. Peel (Chicago: University of Chicago Press, 1972), 123–25.

10. Spencer, *On Social Evolution*, 61–62. Spencer first developed these analogical ideas in his *Social Statics*.

11. Hardy's novels were collected in a series called "Novels of Character and Environment" in 1912. See also H. Grimsditch, *Character and Environment in the Novels of Thomas Hardy* (London: H. F. and G. Witherby, 1925).

12. Indeed, this was very typical of the fiction of the age, whether we consider Thomas Hardy, Gustav Flaubert, Émile Zola, Theodor Fontane, or any number of writers. As Raymond Williams put it, "As in all major realist fiction the quality and destiny of persons and the quality and destiny of a whole way of life are seen in the same dimension and not as separable issues." Williams, *The Country and the City* (London: Chatto and Windus, 1973), 201.

13. Leavis and Thompson, *Culture and Environment*, 1, 93.

14. E. Churchill Semple, *Influences of Geographic Environment* (New York: Holt, 1911); G. Taylor, *The Australian Environment* (Melbourne: Executive Committee of H. A. Hunt, 1918); G. Taylor, *Environment and Race* (Oxford: Oxford University Press, 1927).

15. B. Winterhalder, "Concepts in Historical Ecology: The View from Evolutionary Ecology," in *Historical Ecology: Cultural Knowledge and Chang-

ing Landscapes, ed. C. Crumley (Santa Fe: School of American Research Press, 1994), 28–29.

16. Cited in R.P. McIntosh, *The Background of Ecology: Concept and Theory* (Cambridge: Cambridge University Press, 1985), 40.

17. L. F. Ward, "The Local Distribution of Plants and the Theory of Adaptation," *Popular Science Monthly* 9 (1876): 682.

18. A. Marshall, *The Principles of Economics* (London: Macmillan, 1890).

19. T. Veblen, *Theory of the Leisure Class: An Economic Study in the Evolution of Institution* (New York: Macmillan, 1899); C. L. Morgan, *Habit and Instinct* (London: E. Arnold, 1896), esp. 340; G. Hodgshon, "On the Evolution of Thorstein Veblen's Evolutionary Economics," *Cambridge Journal of Economics* 22 (1998): 415–31.

20. Rev. S. R. Calthrop, "Religion and Science," *Report of the Second Meeting of the National Conference of Unitarian and Other Christian Churches Held in Syracuse, N.Y.* (Boston, 1866), 209.

21. Limits became a main preoccupation for several authors in economic geography and the emerging field of resource economics and conservation. For "peak oil," see M. King Hubbert, "Energy from Fossil Fuels" (paper presented at the Symposium of Energy, Centennial Celebration of the American Association for the Advancement of Science, Washington, DC, September 15, 1948), and published under the same title in *Science* 109 (February 4, 1949): 103–9. Hubbert predicted, very accurately as it turned out (at least before the expansion of shale oil), that American "peak oil" would be reached around 1970. See chapter 3.

22. Douglas R. Weiner, "Russia and the Soviet Union," in *Encyclopedia of World Environmental History*, vol. 3, ed. Shepard Krech III, John R. McNeill, and Carolyn Merchant, 1074–80 (New York: Routledge, 2004).

23. United Nations Educational Scientific and Cultural Organization, "The Scientific Conference on the Conservation and Utilization of Resources," UNSCCUR (a memorandum by the Department of Exact and Natural Sciences, UNESCO), November 10, 1948, 1, UNESCO Archives, Paris.

24. Ibid.

25. IUPN was conceived at a conference in Brunnen, Switzerland, in July 1947 and formally constituted at a meeting in Fontainebleau, France, September 30 to October 7, 1948. It was renamed the International Union for Conservation of Nature (IUCN) in 1956.

26. "Conference for the Establishment of the International Union for the Protection of Nature," UNESCO Archives NS/UIPN/12, Fontainebleau, October 5, 1948, http://unesdoc.unesco.org/images/0015/001547/154724eb.pdf, quotation on 1. See also Roderick Nash, *Wilderness and the American Mind*, 3rd ed. (New Haven, CT: Yale University Press, 1992), 361.

27. Jacques Roger, *Buffon: A Life in Natural History*, ed. L. Pearce Williams, trans. Sarah Lucille Bonnefoi (New York: Columbia University Press, 1997), originally published in French in 1989; Georges-Louis LeClerc, le Compte de Buffon, *The Epochs of Nature*, trans. Jan Zalasiewicz, Anne-Sophie Milon, and Mateusz Zalasiewicz (Chicago: University of Chicago Press, 2018).

28. Alexander von Humboldt, *De distributione geographica plantarum* (Paris: Libraria Graeco-Latina-Germanica, 1817); Michael Dettelbach, "Humboldtian Science," in *Cultures of Natural History*, ed. Nicholas Jardine, James Secord, and Emma Spary (Cambridge: Cambridge University Press, 1996), 287–304; Clarence J. Glacken, *Traces on the Rhodian Shore: Nature and Culture in Western Thought from Ancient Times to the End of the Eighteenth Century* (Berkeley: University of California Press, 1967); Michael Dettelbach, "Global Physics and Aesthetic Empire: Humboldt's Physical Portrait of the Tropics," in *Visions of Empire: Voyages, Botany and Representations of Nature*, ed. David Philip Miller and Peter Hanns Reill, 258–92 (Cambridge: Cambridge University Press, 1996).

29. Robert Millikan, "Alleged Sins of Science," *Scribner's Magazine* (1930), reprinted in *Science and the New Civilization* (New York: Charles Scribner's Sons, 1930), 121.

30. George Evelyn Hutchinson, "On Living in the Biosphere," *Scientific Monthly* 67, no. 6 (1948), 393–97. Hutchinson had already started translations of Vernadsky's work into English at the beginning of the 1940s and used Vernadsky's texts in his teaching at Yale.

31. Iris Borowy, *Defining Sustainable Development for Our Common Future: A History of the World Commission on Environment and Development (Brundtland Commission)* (Abingdon: Routledge, 2014), chap. 2, sec. 2.

32. "Scientists Will Pool Their Knowledge at UNO Conference," *Sydney Morning Herald*, August 19, 1949.

33. In 1964 these lists were to become the famous *Red Lists*, from the color of the field notebooks that the scientists carried. William M. Adams, *Against Extinction: The Story of Conservation* (London: Earthscan, 2004), 130–31.

34. Martin Holdgate, *The Green Web: A Union for World Conservation* (1999; new ed., Abingdon: Earthscan, 2013), 41.

35. Jean-Paul Harroy quoted in Thomas Jundt, *Greening the Red, White, and Blue: The Bomb, Big Business, and Consumer Resistance in Postwar America* (Oxford: Oxford University Press, 2014), 39. Harroy had already demonstrated a capacity for concerned thinking with his 1944 book, *Afrique: Terre qui meurt. La degradation des sols africains sous l'influence de la colonisation* (Brussels: M. Hayez, 1944).

36. Rosalind Irwin, "Posing Global Environmental Problems from Con-

servation to Sustainable Development," in *The International Political Economy of the Environment: Critical Perspectives*, ed. Dimitris Stevis and Valerie J. Assetto (Boulder: Lynne Rienner, 2001), 21–24.

37. Gifford Pinchot's *conservation* ("wise use") had been contrasted with John Muir's *protection* since the turn of the twentieth century. Samuel P. Hays, *Conservation and the Gospel of Efficiency* (1959; new ed., Pittsburgh: University of Pittsburgh Press, 1999).

38. James Rorty, "Hunger Is Obsolete, If—The Unused Weapon to Win the Cold War," *Commentary* (February 1950): 2–3, https://www.commentary magazine.com/articles/hunger-is-obsolete-if-the-unused-weapon-to-win -the-cold-war.

39. Ronald E. Doel, "Constituting the Post-War Earth Sciences: The Military's Influence on the Environmental Sciences in the USA after 1945," *Social Studies of Science* 33 (2003): 635–66. Matthew Farish, "Creating Cold War Climates: The Laboratories of American Globalism," in *Environmental Histories of the Cold War*, ed. J. R. McNeill and Corinna R. Unger (Cambridge: Cambridge University Press, 2010), 51–83. Jacob Darwin Hamblin, *Arming Mother Nature: The Birth of Catastrophic Environmentalism* (Oxford: Oxford University Press, 2013). Simone Turchetti and Peder W. Roberts, eds., *The Surveillance Imperative: Geosciences during the Cold War and Beyond* (London: Palgrave Macmillan, 2014).

40. Cited in H. Nichols, "Greed Held Check to Stretching Natural Resources," *Christian Science Monitor*, September 15, 1948, 9.

41. Harrison Brown, *The Challenge of Man's Future* (New York: Viking, 1954), xi.

42. Ibid., 7.

43. His sources included Roderick Seidenberg, *Post-Historic Man* (Durham: University of North Carolina Press, 1950), and, later, well-known social scientists such as David Riesman (whose bestselling *The Lonely Crowd* he cited), Seymour Martin Lipset, Martin Trow, James Coleman, and Gregory Bateson. He was not alone or the first with big estimates. Statistician George Knibbs, *The Shadow of the World's Future or the Population Possibilities of the Consequences of the Present Rate of Increase of the Earth's Inhabitants* (London: Ernest Benn, 1928), mooted an "absolute limit" of 7.8 billion, also now exceeded.

44. Frank Fraser Darling and John P. Milton, eds., *Future Environments of North America* (Garden City, NY: Natural History Press, 1966).

45. Pierre Dansereau, "Ecological Impact and Human Ecology," in ibid., 425–62.

46. Darling, introduction to Darling and Milton, *Future Environments of North America*, 1–7.

Chapter 3. Resources for Freedom

1. Paul Edwards, *A Vast Machine: Computer Models, Climate Data, and the Politics of Global* Warming (Cambridge, MA: MIT Press, 2010), 366–69. The title of this section is from Donella J. Meadows, John Richardson, and Gerhart Bruckmann, *Groping in the Dark: The First Decade of Global Modeling* (New York: John Wiley, 1982).

2. John McCormick, *The Global Environmental Movement* (London: Belhaven, 1989), 81.

3. Élodie Vielle Blanchard, "Technoscientific Cornucopian Futures versus Doomsday Futures: The World Models and *The Limits to Growth*," in *The Struggle for the Long-Term in Transnational Science and Politics*, ed. Jenny Andersson and Eglė Rindzevičiūtė (Basingstoke: Routledge, 2015), 96; Élodie Vieille Blanchard, "Les limites à la croissance dans un monde global: Modélisations, prospectives, réfutations" (PhD diss., École des Hautes Études en Sciences Sociales, 2011), 351–78; see also Jay Wright Forrester, *Industrial Dynamics* (1961; repr., Eastford, CT: Martino Fine Books, 2013).

4. Blanchard, "Technoscientific Cornucopian Futures," 101.

5. Meadows, Richardson, and Bruckmann, *Groping in the Dark*.

6. Dennis Meadows, Donella Meadows, Jørgen Randers, and William W. Behrens III, *Limits to Growth: A Report for the Club of Rome's Project on the Predicament of Mankind* (London: Universe, 1972), 41, 57, 86, 93.

7. Ibid., 25–33.

8. Ibid., 142.

9. Blanchard, "Technoscientific Cornucopian Futures," 105.

10. Discussed at length in chapter 2.

11. Meadows et al., *Limits to Growth*, 190.

12. Carol S. Carson, "The History of the United States National Income and Product Accounts: The Development of an Analytical Tool," *Review of Income and Wealth* 21 (1975): 153–81; J. Adam Tooze, *Statistics and the German State, 1900–1945: The Making of Modern Economic Knowledge* (Cambridge: Cambridge University Press, 2001), 7–11.

13. Fairfield Osborn, *Limits of the Earth* (London: Faber and Faber, 1954), 11.

14. Samuel Ordway, "Possible Limits of Raw-Material Consumption," in *Man's Role in Changing the Face of the Earth*, ed. William L. Thomas Jr. (Chicago: University of Chicago Press, 1956), 993.

15. J. Frederic Dewhurst et al., *America's Needs and Resources: A Twentieth Century Fund Survey* (New York: Twentieth Century Fund, 1947).

16. Ibid., 676.

17. The President's Materials Policy Commission, *Resources for Freedom* (Washington, DC: Government Printing Office, 1952).

18. Ibid., 1–3.

19. Political and Economic Planning, *World Population and Resources: A Report by PEP* (London: PEP, 1955), xi. This group had been founded at the London School of Economics in 1931, with close links to demographers and ecologists such as Julian Huxley and Alexander Carr-Saunders. Alison Bashford, *Global Population: History, Geopolitics, and Life on Earth* (New York: Columbia University Press, 2014), 168.

20. Osborn, *Limits of the Earth*, 11.

21. H. H. Landsberg, L. L. Fischman, and J. L. Fisher, *Resources in America's Future: Patterns of Requirements and Availabilities, 1960–2000* (Baltimore: Johns Hopkins University Press, 1963).

22. For an indicative article of the time, see George Otis Smith, "Where the World Gets Its Oil, But Where Will Our Children Get It When American Wells Cease to Flow?," *National Geographic* 37 (1920): 181–202. Peter A. Coates, *The Trans-Alaska Pipeline Controversy: Technology, Conservation, and the Frontier* (Fairbanks: University of Alaska Press, 1991), 55; John A. Dugger, "Arctic Oil and Gas: Policy Perspectives," in *United States Arctic Interests, the 1980s and 1990s*, ed. W. E. Westermeyer and K. M. Shusterich, 19–38 (New York: Springer, 1984).

23. Marion King Hubbert, "Energy from Fossil Fuels," *Science* 109 (1949): 103–8.

24. Tyler Priest, "Hubbert's Peak: The Great Debate over the End of Oil," *Historical Studies in the Natural Sciences* 44 (2014): 37–79. Hubbert made several interventions—for example, at the Future Environments of North America conference of 1965 (see chapter 2).

25. Eugene Ayres and C. A. Scarlott, *Energy Sources—the Wealth of the World* (New York: McGraw-Hill, 1952); Hubbert, "Energy from Fossil Fuels," 1949; C. L. Weeks, "The Next Hundred Years Energy Demand and Sources of Supply," *Journal of the Alberta Society of Petroleum Geologists* 9, no. 5 (1961): 141–57; G. A. Lamb, "The Fuel Complex: A Projection," *Annals of the American Academy of Political and Social Science: The Future of Our Natural Resources* 281 (1952): 42–54; Stefan Cihan Aykut, "Energy Futures from the Social Market Economy to the *Energiewende*: The Politicization of West German Energy Debates, 1950–1990," in Andersson and Rindzevičiūtė, *Struggle for the Long-Term*, 68.

26. McCormick, *Global Environmental Movement*, 25.

27. Spencer R. Weart, *The Rise of Nuclear Fear* (Cambridge, MA: Harvard University Press, 2012), 88–95; Roger Williams, *The Nuclear Power Decisions: British Policies, 1953–78* (London: Croom Helm, 1980).

28. Ministry of Fuel and Power, *Fuel and the Future: Proceedings of a Conference, London, 8th–10th October 1946* (London: HMSO, 1948).

29. Martin Chick, *Electricity and Energy Policy in Britain, France and the United States since 1945* (Cheltenham: Edward Elgar, 2007), 7–33.

30. As examples from the sizable literature, see E. Bini and G. Garvini, eds., *Oil Shock: The 1973 Crisis and Its Economic Legacy* (London: I. B. Tauris, 2016); P. Z. Grossman, *U.S. Energy Policy and the Pursuit of Failure* (Cambridge: Cambridge University Press, 2013); Leonardo Maugeri, *The Age of Oil: The Mythology, History, and Future of the World's Most Controversial Resource* (Westport, CT: Praeger, 2006).

31. Robert Thomas Malthus, *An Essay on the Principle of Population* (London: J. Johnson, 1798).

32. Ibid., 3–4.

33. Condorcet cited in ibid., 34, 48.

34. Ibid., 48.

35. Paul Warde, "Fears of Wood Shortage and the Reality of the Woodlands in Europe, c.1450–1850," *History Workshop Journal* 62 (2006): 28–57; Paul Warde, "Early Modern 'Resource Crisis': The Wood Shortage Debates in Europe," in *Crises in Economic and Social History: A Comparative Perspective*, ed. A. T. Brown, Andy Burn, and Rob Doherty, 137–159 (Woodbridge: Boydell, 2015).

36. Cited in Michael Williams, *Deforesting the Earth: From Prehistory to Global Crisis* (Chicago: University of Chicago Press, 2003), 386.

37. Nathaniel S. Shaler, "Earth and Man: An Economic Forecast," *International Quarterly* 10 (1904): 227–39; Nathaniel S. Shaler, "The Exhaustion of the World's Metals," *International Quarterly* 11 (1905): 230–47.

38. Raphael Zon, *The Forest Resources of the World* (Washington, DC: Government Printing Office, 1910); Williams, *Deforesting the Earth*, 390–91; McCormick, *Global Environmental Movement*, 14–17.

39. Rolf Peter Sieferle, *The Subterranean Forest: Energy Systems and the Industrial Revolution* (1982; repr., Cambridge: White Horse Press, 2001).

40. William Stanley Jevons, *The Coal Question: An Enquiry concerning the Progress of the Nation, and the Probable Exhaustion of Our Coal Mines* (London: Macmillan, 1865).

41. Margaret Schabas, *A World Ruled by Number: William Stanley Jevons and the Rise of Mathematical Economics* (Princeton, NJ: Princeton University Press, 1990).

42. *Report of the Commissioners Appointed to Inquire into the Several Matters Relating to Coal in the United Kingdom* (London: HMSO, 1871); Political and Economic Planning, *World Population and Resources*, 70.

43. Matthew J. Connelly, *Fatal Misconception: The Struggle to Control World Population* (Cambridge, MA: Harvard University Press, 2008); Matthew Connelly, "Population Control in India: Prologue to the Emergency

Period," *Population and Development Review* 32 (2006): 629–67; Alison Bashford, "Nation, Empire, Globe: The Spaces of Population Debate in the Interwar Years," *Comparative Studies in Society and History* 49 (2007): 170–201; Alison Bashford, *Global Population: History, Geopolitics and Life on Earth* (New York: Columbia University Press, 2014).

44. Bashford, *Global Population*, 82–87.

45. Pearl and Reed used census data to try and fix growth to the logistic curves they developed in studying fruit fly populations. The approach was much debated and controversial. Putnam assessed their predictions in 1953, finding the error for the United States to be very small (0.3 percent) but in regard to the Philippines, very large (122 percent), and found an underestimate of 27 percent for the whole world. Raymond Pearl and Lowell J. Reed, "On the Rate of Growth of the Population of the United States since 1790 and Its Mathematical Representation," *Proceedings of the National Academy of Sciences of America* 6 (1920): 275–88. Palmer C. Putnam, *Energy in the Future* (New York: Van Nostrand, 1953), 40. See also Sharon E. Kingsland, *Episodes in the History of Population Ecology* (Chicago: University of Chicago Press, 1985); Bashford, *Global Population*, 81, 88–89; Edmund Ramsden, "Carving Up Population Science: Eugenics, Demography and the Controversy over the 'Biological Law' of Population Growth," *Social Studies of Science* 32 (2002): 857–99.

46. Alexander M. Carr-Saunders, *The Population Problem: A Study in Human Evolution* (Oxford: Clarendon, 1922); Charles Elton, *Animal Ecology* (London: Sidgwick and Jackson, 1927).

47. Peder Anker, *Imperial Ecology: Environmental Order in the British Empire, 1895–1945* (Cambridge MA; Harvard University Press, 2001), 86–93; Bashford, *Global Population*, 91–94, 159–61.

48. Thomas Robertson, *The Malthusian Moment: Global Population and the Birth of American Environmentalism* (New Brunswick: Rutgers University Press, 2012), 16–23; Kingsland, *Episodes*; Bashford, *Global Population*, 197, 202–6; Greg Mitman, *The State of Nature: Ecology, Community, and American Social Thought, 1900–1950* (Chicago: University of Chicago Press, 1992), 89.

49. Ariane Tanner, *Die Mathematisierung des Lebens: Alfred James Lotka und der energetische Holismus im 20. Jahrhundert* (Tübingen, Mohr Siebeck, 2017); Kingsland, *Episodes*, 23–26.

50. Ariane Tanner, "Publish and Perish: Alfred James Lotka und die Anspannung in der Wissenschaft," *NTM Zeitschrift für Geschichte der Wissenschaften, Technik und Medizin* 21 (2012): 143–70.

51. Ibid., 102–3, 108–9. Alfred J. Lotka, "Contribution to the Energetics of Evolution," *Proceedings of the National Academy of Sciences of the USA* 8 (1922): 147–51; Alfred J. Lotka, *Elements of Physical Biology* (Baltimore:

Williams and Watkins, 1925). For another similar contemporary work, see Royal N. Chapman, "The Quantitative Analysis of Environmental Factors," *Ecology* 9 (1928): 111–22.

52. Eugene P. Odum, in collaboration with Howard T. Odum, *Fundamentals of Ecology* (Philadelphia: W. S. Saunders, 1953).

53. Hubbert, "Energy from Fossil Fuels"; see also Putnam, *Energy in the Future*; Harrison Brown, *The Challenge of Man's Future* (New York: Viking, 1954).

54. Bashford, *Global Population*, 55–106. For an overview of population estimates, see also Paul Demeny, "Demography and the Limits to Growth," *Population and Development Review* 14, supplement (1988): 213–44.

55. Björn-Ola Linnér, *The Return of Malthus: Environmentalism and Post-War Population-Resource Crisis* (Isle of Harris, UK: White Horse Press, 2003).

56. Georg Borgström, *The Hungry Planet: The Modern World at the Edge of Famine* (New York: Macmillan, 1965); originally published in Swedish in 1953. National and cultural styles are embedded in discourses; American prophets of doom seem more concerned with freedom, while Borgström, the Swede, was more concerned with global equity.

57. In a 1976 interview he calls himself a "realist." "When a house is on fire, the optimist says the fire will go out by itself, the pessimist says it is no use trying to stop the fire, whereas the realist says 'Let's stop the fire!'" Tord Hubert, "Georg Borgström varnar den rika världen: Vi har bara tio år på oss!," *Vecko-Journalen* 44 (1976): 15.

58. Bashford, *Global Population*, 292–93.

59. Borgström, *Hungry Planet*. For contemporary ideas on desertification, see Paul Sears, *Deserts on the March*, 2nd ed. (London: Routledge and Keegan Paul, 1949). In 1951, at India's initiative, UNESCO launched an Advisory Committee on Arid Zone Research, specifically to work on desertification. Questions about rising sea levels became part of the early global warming discourse in the 1950s. See Sverker Sörlin, "The Global Warming That Did Not Happen: Historicizing Glaciology and Climate Change," in *Nature's End: History and the Environment*, ed. Paul Warde and Sverker Sörlin, 93–114 (Basingstoke: Palgrave Macmillan, 2009).

60. Paul R. Ehrlich, *The Population Bomb: Population Control or the Race to Oblivion* (New York, Ballantine, 1968). As Bashford notes, the metaphor of population growth as a "bomb" had already been employed before the nuclear age by the key theorist of the demographic transition, Kingsley Davis, in 1944. Bashford, *Global Population*, 306.

61. Paul Sabin, *The Bet: Paul Ehrlich, Julian Simon, and Our Gamble over Earth's Future* (New Haven, CT: Yale University Press, 2013), 2–3. Douglas MacEwan, one of the founders of the Malthusian Conservation Society in

the United Kingdom, credited his inspiration to an article by Julian Huxley in *Playboy*. Horace Herring, "The Conservation Society: Harbinger of the 1970s Environment Movement in the UK," *Environment and History* 7 (2001): 386.

62. See Sabin, *The Bet*, 10–12, 20; Thomas Robertson, "Total War and the Total Environment: Fairfield Osborn, William Vogt and the Birth of Global Ecology," *Environmental History* 17 (2012): 336–64.

63. Fabien Locher, "Les pâturages de la Guerre froide: Garrett Hardin et la 'Tragédie des communs,'" *Revue d'Histoire Moderne et Contemporaine* 60 (2013): 7–36. For a longer history of these issues connecting Aldo Leopold, William Vogt, and others, see Miles A. Powell, "'Pestered with inhabitants': Aldo Leopold, William Vogt, and More Trouble with Wilderness," *Pacific Historical Review* 84 (2015): 195–226.

64. Sabine Höhler, *Spaceship Earth in the Environmental Age, 1960–1990* (New York: Berghahn, 2015); Richard Buckminster Fuller, *Operating Manual for Spaceship Earth* (Carbondale: Southern Illinois University Press, 1968).

65. Ehrlich, *Population Bomb*; Georg Borgström, *Too Many: A Biological Overview of the Earth's Limitations* (London: Macmillan, 1969).

66. Sabin, *The Bet*, 87–93; Forrester was characterized as a "brash engineer" by the decision theorist Hebert Simon. Eglė Rindzevičiūtė, *The Power of Systems: How Policy Sciences Opened Up the Cold War World* (Ithaca, NY: Cornell University Press, 2016), 58; McCormick, *Global Environmental Movement*, 81–83.

67. Zuckerman cited in Rindzevičiūtė, *Power of Systems*, 135.

68. For an overview of postwar growth theory, see Flavio Comim, "On the Concept of Applied Economics: Lessons from Cambridge Economists and the History of Growth Theories," *History of Political Economy* 132, Winter supplement (2000): 147–76; and E. Roy Weintraub, ed., "MIT and the Transformation of American Economics," *Journal of Political Economy* 46, supplement 1 (2014).

69. An accessible and wide-ranging discussion of these debates within economics is provided in Roger E. Backhouse, *The Puzzle of Modern Economics: Science or Ideology?* (Cambridge: Cambridge University Press, 2010).

70. Robert M. Solow, "A Contribution to the Theory of Economic Growth," *Quarterly Journal of Economics* 70 (1956): 65–94; Robert M. Solow, "Technical Change and the Aggregate Production Function," *Review of Economics and Statistics* 39 (1957): 312–30.

71. Robert M. Solow, "The Economics of Resources or the Resources of Economics," *American Economic Review* 64 (1974): 1–14; Sabin, *The Bet*, 53.

72. Brown, *Challenge of Man's Future*, 3–45.

73. Morris A. Adelman, "Economic Background: Impact on Markets and Trading Patterns," in *Le pétrole et le gaz Artiques*, ed. Jean Malaurie (Le Havre: Colloque International de la Fondation Francaise d'Etudes Nordiques, May 2–5 1973), 80.

74. Harold J. Barnett and Chandler Morse, *Scarcity and Growth: The Economics of Natural Resource Availability* (Baltimore: Johns Hopkins University Press for Resources for the Future, 1963).

75. See William Nordhaus, "World Dynamics: Measurement without Data," *Economic Journal* 83, no. 332 (1973); and William Nordhaus, "World Modelling from the Bottom Up," IIASA Research Memorandum, March 1975.

76. Resources for the Future Staff, "FEEM 20th Anniversary Prize in Environmental Economics," June 30, 2010, http://www.rff.org/blog/2010 /feem-20th-anniversary-prize-environmental-economics. FEEM is an Italian think tank established by the Italian national oil company (ENI) and named after their once director, specializing in environmental economics and with a major international reputation.

77. Harold Hotelling, "The Economics of Exhaustible Resources," *Journal of Political Economy* 39 (1931): 137–75.

78. Grossman, *U.S. Energy Policy*.

79. Gerald O. Barney, *The Global 2000 Report to the President of the United States, Entering the Twenty-First Century* (Washington, DC: Government Printing Office, 1980). A summary of the report is provided by Gus Speth, "The Global 2000 Report to the President," *Boston College Environmental Affairs Law Review* 8 (1980): 695–703.

80. Sabin, *The Bet*; Julian L. Simon and Herman Kahn, eds., *The Resourceful Earth: A Response to Global 2000* (Oxford: Blackwell, 1984).

81. Sabin, *The Bet*.

82. Ibid., 203.

83. For a broad overview, see Michael Common and Sigrid Stagl, *Ecological Economics: An Introduction* (Cambridge: Cambridge University Press, 2005).

Chapter 4. Ecology on the March

1. John Steinbeck, *The Grapes of Wrath* (1939; repr., London: Penguin, 1992), 92.

2. Francis Ratcliffe, *Flying Fox and Drifting Sand: The Adventures of a Biologist in Australia* (1938; repr., Sydney: Angus and Robertson, 1947), 332.

3. D. W. Meinig, *On the Margins of the Good Earth: The South Australian Wheat Frontier 1869–1884* (Adelaide: Rigby, 1970); Libby Robin, *How a Continent Created a Nation* (Sydney: University of New South Wales Press, 2007).

4. Paul Sears, *Deserts on the March*, 2nd ed. (London: Routledge and Kegan Paul, 1949), 177.

5. The first edition of *Deserts on the March* was published in 1935 by the University of Oklahoma Press. The US Soil Conservation Service, established in 1935, continues today as NRCS (Natural Resources Conservation Service). Soil conservation authorities were established in New South Wales in 1938 and in Victoria in 1940.

6. Sears, *Deserts on the March*, 176.

7. Ibid., 168.

8. Cameron Muir, *The Broken Promise of Agricultural Progress* (London: Routledge, 2014).

9. Elyne Mitchell, *Soil and Civilization* (Sydney: Angus and Robertson, 1946).

10. David Moon, "The Environmental History of the Russian Steppes," *Transactions of the Royal Historical Society* 15 (2005): 149–74; see also David Moon, *The Plough That Broke the Steppes: Agriculture and Environment on Russia's Grasslands, 1700–1914* (New York: Oxford University Press, 2013).

11. Libby Robin, "'Ecology: A Science of Empire?,'" in *Ecology and Empire: Environmental History of Settler Societies*, ed. Tom Griffiths and Libby Robin, 63–75 (Keele, UK: Edinburgh University Press, 1997).

12. Mitchell, *Soil and Civilization*, 1.

13. Ibid., 2.

14. Ibid., 53, 1, 3, respectively.

15. Aldo Leopold, "The Land Ethic," in *A Sand County Almanac*, ed. Robert Finch, 201–28 (1949; repr., New York: Oxford University Press, 1987), 210.

16. Cameron Muir, "Feeding the World: Our Great Myth," *Griffith Review* 27 (2010): 59–73; Alison Bashford, *Global Population: History, Geopolitics and Life on Earth* (New York: Columbia University Press, 2014).

17. Borgström's work on ghost acreage first appeared in Swedish in 1953. In 1962, his book *The Hungry Planet: The Modern World at the Edge of Famine* was translated into English. The chapter on ghost acreage is republished in Libby Robin, Sverker Sörlin, and Paul Warde, eds., *The Future of Nature: Documents of Global Change* (New Haven, CT: Yale University Press, 2013), 40–50.

18. Bashford, *Global Population*.

19. Carmel Finley, *All the Fish in the Sea: Maximum Sustainable Yield and the Failure of Fisheries* (Chicago: University of Chicago Press, 2011); D. Graham Burnett, *The Sounding of the Whale: Science and Cetaceans in the Twentieth Century* (Chicago: University of Chicago Press, 2012).

20. This translation from German is cited in Donald Worster, *Nature's*

Economy: A History of Ecological Ideas (1977; repr., New York: Cambridge University Press, 1985), 192.

21. Eugene Cittadino, "Ecology and the Professionalization of Botany in America, 1890–1905," *Studies in the History of Biology* 4 (1980): 171–98; Ronald Tobey, "Theoretical Science and Technology in American Ecology," *Technology and Culture* 17 (1976): 718–28; Richard A. Overfield, "Charles E. Bessey: The Impact of the 'New' Botany on American Agriculture, 1880–1910," *Technology and Culture* 16 (1975): 162–81.

22. Worster, *Nature's Economy*, 203.

23. Comparison of climates at different latitudes and altitudes dated from the "biogeography" of Alexander von Humboldt, who developed its principles through his important travels in the early 1800s. See A. von Humboldt and A. Bonpland, *Essay on the Geography of Plants*, trans. S. Romanowski, ed. S. T. Jackson with accompanying essays and supplementary material by S. T. Jackson and S. Romanowski (1807; repr., Chicago: University of Chicago Press, 2009).

24. R. J. Goodland, "The Tropical Origin of Ecology: Eugen Warming's Jubilee," *Oikos* 26 (1976): 240.

25. A. G. Tansley, "The Early History of Modern Plant Ecology in Britain," *Journal of Ecology* 35 (1947): 130.

26. Ronald Tobey, *Saving the Prairies: The Life Cycle of the Founding School of American Plant Ecology, 1895–1955* (Berkeley: University of California Press, 1981).

27. William Hoffman, "The Tallgrass Prairie: Raymond Lindeman, a Minnesota Bog Lake and the Birth of Ecosystems Ecology," in *Conservation on the Northern Plains: New Perspectives*, ed. A. J. Amato, 1–31 (Sioux Falls, SD: Center for Western Studies, Augustana University, 2017).

28. Frederic Clements, *Plant Succession: An Analysis of the Development of Vegetation* (Washington, DC: Carnegie Institution of Washington, 1916), publication no. 242; Roscoe Pound and Frederic Clements, *Phytogeography of Nebraska* (1898; rev. ed. 1900; repr., New York: Arno Press, 1977).

29. Gregg Mitman, *The State of Nature* (Chicago: University of Chicago Press, 1992), 38–44.

30. Charles Elton, *Animal Ecology*, with new introductory material by Matthew A. Leibold and Timothy J. Wootton (1926; repr., Chicago: University of Chicago Press, 2001).

31. Charles Elton, preface to *Animal Ecology* (London: Sidgwick and Jackson, 1927). William Vogt (see chapter 1) took a copy of Elton's book with him on his South American research journeys.

32. In the end, Shelford transferred his considerable personal energies to conservation, setting up an activist organization that became the Nature

Conservancy, one of the major conservation nongovernmental organizations (NGOs) in the United States and now the international Nature Conservancy (http://www.nature.org). Note that this is not the same thing as the Nature Conservancy in Britain, established in 1949.

33. Peter Crowcroft, *Elton's Ecologists: A History of the Bureau of Animal Population* (Chicago: University of Chicago Press, 1991).

34. Libby Robin, "Ecology: A Science of Empire?," in *Ecology and Empire: Environmental History of Settler Societies,* ed. Tom Griffiths and Libby Robin, 63–75 (Keele, UK: Edinburgh University Press, 1997); Michael Worboys, "Science and British Colonial Imperialism, 1895–1940" (PhD thesis, University of Sussex, 1979), 305.

35. Francis Ratcliffe's *Flying Fox and Drifting Sand* (1947) was prescribed as reading in schools throughout the 1950s, and even as part of a Cold War disinformation campaign in Asia designed to inform people of the unsuitability of Australia for settlement. Libby Robin and Tom Griffiths, "Francis Noble Ratcliffe, 1904–1970," in *New Dictionary of Scientific Biography,* vol. 6, ed. Noretta Koertge, 207–11 (Farmington Hills, MI: Charles Scribner's Sons, 2007).

36. Frank Golley, *A History of the Ecosystem Concept in Ecology: More Than the Sum of the Parts* (New Haven, CT: Yale University Press, 1993).

37. R. N. Chapman, "The Quantitative Analysis of Environmental Factors," *Ecology* 9 (1928): 111–22.

38. Arthur Tansley, "The Use and Abuse of Vegetational Concepts and Terms," *Ecology* 16 (1935): 284–307. Tansley in his early years had also worked as a researcher for Herbert Spencer.

39. Formal exchanges began with the International Phytogeographic Excursion (IPE) to the Norfolk Broads in England in 1911. See Laura Cameron and David Matless, "Translocal Ecologies: The Norfolk Broads, the 'Natural,' and the International Phytogeographical [sic] Excursion, 1911," *Journal of the History of Biology* 44 (2011): 15–41.

40. The Swiss botanist Josias Braun-Blanquet published *Plant Sociology: The Study of Plant Communities* (New York: Hafner, 1972). Originally published in 1964 in German as *Pflanzensoziologie.* Braun-Blanquet established the SIGMATISTs (later called the Braun-Banquet school of vegetation analysis), which also included Carl Schröter in Zurich, who was particularly influential in Britain. Robert P. McIntosh, *The Background of Ecology* (Cambridge: Cambridge University Press, 1985), 40–44.

41. John Phillips, "Succession, Development, the Climax and the Complex Organism: An Analysis of Concepts," parts 1–3, *Journal of Ecology* 22 (1934): 554–71; 23 (1935): 210–46, 488–508, respectively. See also Frederic E. Clements, "Nature and Structure of the Climax," *Journal of Ecology* 24, no. 1

(1936): 252–84, a reply rich in the prairie story. Tansley placed his paper in the American journal *Ecology*, while Clements published in the British journal.

42. Jan Smuts was a leader and architect of the independent and racist South African state, as well as a senior military commander in both world wars, with a keen interest in ecology.

43. Peder Anker, *Imperial Ecology: Environmental Order in the British Empire* (Cambridge, MA: Harvard University Press, 2001), 124–35.

44. Laura Cameron, "Sir Arthur George Tansley," in *New Dictionary of Scientific Biography*, vol. 7, ed. Noretta Koertge, 3–9 (Farmington Hills, MI: Charles Scribner's Sons, 2007).

45. Tansley, "Use and Abuse," 340.

46. Ibid., 289.

47. Arnold G. Van der Valk, "From Formation to Ecosystem: Tansley's Response to Clements' Climax, "*Journal of the History of Biology* 47 (2014): 293–321.

48. Bernard C. Patten and Eugene P. Odum, "The Cybernetic Nature of Ecosystems," *American Naturalist* 118, no. 6 (December 1981): 886–95.

49. H. G. Wells, Julian Huxley, and G. P. Wells, *The Science of Life* (London: Cassell, 1931), 578.

50. Charles Elton, *The Ecology of Animals* (London: Methuen, 1933). Vito Volterra, "Fluctuations in the Abundance of a Species Considered Mathematically," *Nature* 118 (October 16, 1926): 558–60. A. J. Lotka, *Elements of Physical Biology* (Baltimore: Williams and Wilkins, 1925). The argument about the influence of Lotka and Volterra on Elton and Huxley is discussed by Warwick Anderson, "Postcolonial Ecologies of Parasite and Host: Making Parasitism Cosmopolitan," *Journal of the History of Biology*, 49, no. 2 (2016): 241–59. See also Sharon E. Kingsland, *Modeling Nature: Episodes in the History of Population Ecology* (Chicago: University of Chicago Press, 1985).

51. Clements also modified his original climax theory in the 1930s to incorporate more emphasis on adapting to climatic changes (such as drought), rather than on the perfect reproduction of nature's original "biome" or ideal (climax) plant and animal community. Frederic E. Clements, "Nature and Structure of the Climax," *Journal of Ecology* 24, no. 1 (1936): 252–84.

52. Daniel C. Coleman, *Big Ecology: The Emergence of Ecosystem Science* (Berkeley, CA: University of California Press, 2010).

53. As expressed by Turner in an interview at his home in Castlemaine, Australia, with Libby Robin, August 28, 1990.

54. The next major fires in the same area in 2009, also disastrous, were also caused by the same combination of climate and fuel, regrown after seventy years. Tom Griffiths, "We Have Still Not Lived Long Enough," *In-*

side Story, February 16, 2009, http://inside.org.au/we-have-still-not-lived
-long-enough.

55. Interview with Libby Robin, January 28, 1991.

56. Tom Griffiths, *Forests of Ash: An Environmental History* (Cambridge: Cambridge University Press, 2002).

57. Rhys Jones, "Fire-stick Farming," *Australian Natural History* 16 (1969): 224; on the Malakunanja II site, see R. G. Roberts, Rhys Jones, and M. A. Smith, "Thermoluminescence Dating of a 50,000 Year Human Occupation Site in Northern Australia," *Nature* 345 (1990): 153–56.

58. Ian Tyrrell, *True Gardens of the Gods: Californian-Australian Environmental Reform, 1860–1930* (Berkeley: University of California Press, 1999); Jared Farmer, *Trees in Paradise: A California History* (New York: W. W. Norton, 2013); J. Carruthers, L. Robin, J. Hattingh, C. Kull, H. Rangan, and B. W. van Wilgen, "A Native at Home and Abroad: The History, Politics, Ethics and Aesthetics of *Acacia*," *Diversity and Distributions* 17, no. 5 (September 2011): 810–21.

59. Alfred Crosby, *Ecological Imperialism: The Biological Expansion of Europe, 900–1900* (Cambridge: Cambridge University Press, 1986).

60. Charles Elton, *The Ecology of Invasions by Animals and Plants* (Chicago: University of Chicago Press, 1958), 33.

61. SCOPE worked under the auspices of the International Council of Scientific Unions, later renamed the International Council for Science (ICSU), one of the oldest nongovernment organizations in the world, founded in 1931 to promote international scientific activity in the different branches of science and its application for the benefit of humanity. Laura Huenneke, Dennis Glick, F. W. Waweru, Robert L. Brownell Jr., and R. Goodland, "SCOPE Program on Biological Invasions: A Status Report," *Conservation Biology* 2, no. 1 (1988): 8–10.

62. David M. Richardson, ed., *Fifty Years of Invasion Ecology: The Legacy of Charles Elton* (Oxford: Wiley-Blackwell, 2010); Mark A. Davis, *Invasion Biology* (Oxford: Oxford University Press, 2009).

63. Quoted in Emma Marris, *Rambunctious Garden: Saving Nature in a Post-Wild World* (New York: Bloomsbury, 2011), 30. See also Daniel Botkin, *The Moon in the Nautilus Shell: Discordant Harmonies Reconsidered* (Oxford: Oxford University Press, 2012); Daniel Botkin, *Discordant Harmonies: A New Ecology for the Twenty-First Century* (Oxford: Oxford University Press, 1990).

64. E. P. Odum (in collaboration with H. T. Odum), *Fundamentals of Ecology* (Philadelphia: W. S. Saunders, 1953), 9; emphasis added. Further discussed in the final section of this chapter.

65. On Big Science, see Derek J. de Solla Price, *Little Science, Big Science* (1963; repr., New York: Columbia University Press, 1986).

66. Golley, *History of the Ecosystem*.

67. Raymond L. Lindeman, "The Trophic-Dynamic Aspect of Ecology," *Ecology* 23 (1942): 399–417. G. Evelyn Hutchinson, "On Living in the Biosphere," *Scientific Monthly* 67 (1948): 393–97. (See also chapter 5 on Hutchinson and Vladimir Vernadsky.) Robert McIntosh, *The Background to Ecology* (Cambridge: Cambridge University Press, 1986), 196–99.

68. Coleman, *Big Ecology*; Odum, *Fundamentals of Ecology*; Stephen Bocking, "Commentary on Odum's *Fundamentals of Ecology*," in *The Future of Nature*, ed. Libby Robin, Sverker Sörlin, and Paul Warde, 242–44 (New Haven, CT: Yale University Press, 2013). Bocking writes at length about Odum's work at Oak Ridge in *Ecologists and Environmental Politics: A History of Contemporary Ecology* (New Haven, CT: Yale University Press, 1997).

69. Coleman, *Big Ecology*, 4.

70. Also called Lotka-Volterra equations, predator-prey equations are a pair of first-order, nonlinear, differential equations describing a dynamic biological system in which two species (one a predator and one its prey) interact and sometimes co-evolve. Lotka trained Raymond Pearl (see chapter 3).

71. The IBP was international and stimulated at first by the example of the International Geophysical Year of 1957 (see chapter 5), but was also the first of the NSF's programs, which continue up until the present. The NSF also funds the Long-Term Ecological Research Network (LTERN), which is significant in both North and South America. In Australia, it is also important, but funded differently. See David Lindenmayer, Emma Burns, Nicole Thurgate, and Andrew Lowe, eds., *Biodiversity and Environmental Change: Monitoring, Challenges and Direction* (Melbourne: CSIRO, 2014).

72. Golley, *History of the Ecosystem*, 2.

73. Ioan Fazey, Joern Fischer, and David Lindenmayer, "What Do Conservation Biologists Publish?," *Biological Conservation* 124 (2005): 63–73. Ioan Fazey, Joern Fischer, and David Lindenmayer, "Who Does All the Research in Conservation Biology?," *Biodiversity and Conservation* 14 (2005): 917–34.

74. Timothy Farnham, *Saving Nature's Legacy: The Origins of the Idea of Biodiversity* (New Haven, CT: Yale University Press, 2007); Libby Robin, "The Rise of the Idea of Biodiversity: Crises, Responses and Expertise," *Quaderni (Journal of l'Institut des Sciences Humaines et Sociales du CNRS)*, special issue, "Les promesses de la biodiversité," 76, no. 1 (2011): 25–38. On crisis, see Michael Soulé, "What Is Conservation Biology?," *Bioscience* 35, no. 11 (1985): 727–34.

75. Conservation International (1998), "Ólafur Ragnar Grimsson on Climate Change," http://www.conservation.org/documentaries/Pages/megadiversity.aspx (accessed February 16, 2018).

76. Soulé, "What Is Conservation Biology?"

77. See chapter 1.

78. Gretchen Daily, *Nature's Services: Societal Dependence on Natural Eco-systems* (Washington, DC: Island Press, 1997); W. E. Westman, "How Much Are Nature's Services Worth?," *Science* 197, no. 4307 (September 1977): 960–64; Richard B. Norgaard, "Commentary on Daily *Nature's Services*," in Robin, Sörlin, and Warde, *Future of Nature*, 462–64.

79. Anthony D. Barnosky et al., "Has the Earth's Sixth Mass Extinction Already Arrived?," *Nature* 471 (March 3, 2011): 51–57.

80. Wallace S. Broecker, "Unpleasant Surprises in the Greenhouse?," *Nature* 328 (July 9, 1987): 123–26.

81. Johan Rockström, Will Steffen et al., "Planetary Boundaries: Exploring the Safe Operating Space for Humanity," *Ecology and Society* 14, no. 2 (2009): 32, http://www.ecologyandsociety.org/vol14/iss2/art32, fig 1.

82. Sandra Diaz, Sebsebe Demissew, Carlos Joly, W. Mark Lonsdale, and Anne Larigauderie, "A Rosetta Stone for Nature's Benefits to People," *PLOS Biology* 13, no. 1 (2015); Richard Norgaard, "Finding Hope in the Millennium Assessment," *Conservation Biology* 22, no. 4 (2008): 862–69. The Millennium Ecosystem Assessment is commonly called the MA.

83. McIntosh, *Background to Ecology*, 203.

Chapter 5. Climate Enters the Environment

1. Wallace S. Broecker, "Unpleasant Surprises in the Greenhouse?," *Nature* 328 (1987): 123.

2. Ibid., 124.

3. J. Petit et al., "Climate and Atmospheric History of the Past 420,000 Years from the Vostok Ice Core, Antarctica," *Nature* 399 (1999): 429–36. Daniel B. Botkin, *Discordant Harmonies: A New Ecology for the Twenty-First Century* (New York: Oxford University Press, 1990); Niles Eldredge and Stephen Jay Gould, "Punctuated Equilibria: An Alternative to Phyletic Gradualism," in *Models in Paleobiology*, ed. T. J. M. Schopf, 82–115 (San Francisco: W. H. Freeman, 1972); Tom Griffiths, "Commentary," in *The Future of Nature: Documents of Global Change*, ed. Libby Robin, Sverker Sörlin, and Paul Warde (New Haven, CT: Yale University Press, 2013), 359–62.

4. Naomi Klein, *This Changes Everything: Climate vs. Capitalism* (New York: Norton, 2014), makes this point decisively, but it is the tip of the iceberg of literature on the same topic.

5. Mike Hulme, "Reducing the Future to Climate: A Story of Climate Determinism and Reductionism," *Osiris* 26 (2011), ed. James Rodger Fleming and Vladimir Jankovich, 245–66.

6. Robert E. Johnson and Robert M. Kark, "Feeding Problems in Man as

Related to Environment: An Analysis of United States and Canadian Army Ration Trials and Surveys, 1941–1946," June 30, 1946, mimeograph, Harvard Fatigue Laboratory collections, Box 24, Folder 13, Harvard Medical School Archives, Boston. In this and other reports from the Harvard Fatigue Lab, the concept "environment" was not problematized. It had the obvious meaning of a site, climate, or geographical space or region. For general information on the institution, see Steven Horvath and Elizabeth Horvath, *The Harvard Fatigue Laboratory: Its History and Contributions* (Englewood Cliffs, NJ: Prentice-Hall, 1973). The laboratory's director, Bruce Dill, summarized the research tradition he had in many respects founded in *Life, Heat, and Altitude: Physiological Effects of Hot Climates and Great Heights* (Cambridge, MA: Harvard University Press, 1938). For the persistent use in this tradition of environment as a stress factor, see Steven M. Horvath and Mohamed K. Yousef, eds., *Environmental Physiology: Aging, Heat and Altitude* (New York: Elsevier/North-Holland, 1981); and B. M. Marriott and S. J. Carlson, eds., *Nutritional Needs in Cold and in High-Altitude Environments: Applications for Military Personnel in Field Operations* (Washington, DC: National Academic Press, 1996). For details on the Fatigue Lab's methodology and research style, see Andi Johnson, " 'They Sweat for Science': The Harvard Fatigue Laboratory and Self-Experimentation in American Exercise Physiology," *Journal of the History of Biology* 48, no. 3 (2015): 425–54.

7. President's Science Advisory Committee, *Restoring the Quality of Our Environment: Report of the Environmental Pollution Panel* (Washington, DC: Government Printing Office, 1965), 2.

8. Nico Stehr and Hans von Storch, eds., introduction to *Eduard Brückner—The Sources and Consequences of Climate Change and Climate Variability in Historic Times* (Dordrecht: Springer Science-Business Media, 2000), esp. 12.

9. U. Radok, "The International Commission on Snow and Ice (ICSI) and Its Precursors, 1894–1994," *Hydrological Sciences Journal* 42, no. 2 (1997): 131–40.

10. Homer Eugene LeGrand, *Drifting Continents and Shifting Theories: The Modern Revolution in* Geology *and Scientific Change* http://catalogue.nla .gov.au/Search/Home?lookfor=author:%22LeGrand,%20H.%20E.%20 (Homer%20Eugene),%201944-%22&iknowwhatimean=1 (Cambridge: Cambridge University Press, 1988).

11. Helen Rozwadowski, *Fathoming the Ocean: The Discovery and Exploration of the Deep Sea* (Cambridge, MA: Harvard University Press, 2005); James R. Fleming, *Inventing Atmospheric Science: Bjerknes, Rossby, Wexler, and the Foundations of Modern Meteorology* (Cambridge, MA: MIT Press, 2016); Jeremy Namias, "The History of Polar Front and Air Mass Concepts in the

United States—An Eyewitness Account," *Bulletin of the American Meteorological Society* 64 (1983): 734–55.

12. Paul A. Edwards, *A Vast Machine: Computer Models, Climate Data, and the Politics of Global Warming* (Cambridge, MA: MIT Press, 2010).

13. George Perkins Marsh, *Man and Nature: or, Physical Geography as Modified by Human Action* (New York: Scribner, 1864); David Lowenthal, *George Perkins Marsh: Versatile Vermonter* (1958; rev. ed. Seattle: University of Washington Press, 2009).

14. Vladimir Vernadsky, *La géochimie* [Geochemistry] (Paris: Librairie Félix Alcan, 1924).

15. Originally published in 1926 in Russian, translated into French in 1929, but only much later into English, when the concept had already been adopted by UNESCO and started a global career. Vladimir I. Vernadsky, *The Biosphere*, ed. Mark McMenamin (New York: Copernicus, 1998); K. E. Bailes, *Science and Russian Culture in an Age of Revolutions: V. I. Vernadsky and His Scientific School, 1863–1945* (Bloomington: Indiana University Press, 1990). Jacques Grinevald, "Introduction: The Invisibility of the Vernadskian Revolution," in Vernadsky, *Biosphere*, 20–32. Andrey V. Lapo, *Traces of Bygone Biospheres* (Moscow: Mir Publishers, 1987); Jonathan Oldfield and D. J. B. Shaw, "V. I. Vernadsky and the Noosphere Concept: Russian Understandings of Society-Nature Interaction," *Geoforum* 37, no. 1 (2006): 145–54; Jonathan Oldfield and D. J. B. Shaw, "V. I. Vernadskii and the Development of Biogeochemical Understandings of the Biosphere, c.1880s–1968," *British Journal for the History of Science* 46, no. 2 (2013): 287–310. Pey-Yi Chu, "Vladimir I. Vernadsky, *The Biosphere* (1926): Excerpts and Commentary," in Robin, Sörlin, and Warde, *Future of Nature*, 161–73.

16. G. Evelyn Hutchinson, "On Living in the Biosphere," *Scientific Monthly* 67 (1948); Vladimir Vernadsky, "The Biosphere and the Noosphere," *American Scientist* 33 (1945): 1–12.

17. Svante Arrhenius, "On the Influence of Carbonic Acid in the Air upon the Temperature of the Ground," *Philosophical Magazine* 41 (1896): 237–76.

18. G. Evelyn Hutchinson, "The Biosphere," *Scientific American* 223, no. 3 (1970): 45–53; reprinted in *The Biosphere (A Scientific American Book)* (San Francisco: W. H. Freeman, 1970), 3–11.

19. Arthur Eddington, *The Nature of the Physical World* (Cambridge: Cambridge University Press, 1928), 310–11.

20. Many, however, worked on radiative transfer that sometimes was directly linked to the greenhouse effect; see, for example, Frank Very, "The Greenhouse Theory and Planetary Temperatures," *Philosophical Magazine* 16 (1908): 462–80. Later, in the 1940s, it was more commonly studied be-

cause radiative heating and cooling rates were important for weather prediction. See David Archer and Raymond Pierrehumbert, *The Warming Papers: The Scientific Foundation for the Climate Change Forecast* (Oxford: Wiley, 2011).

21. See chapter 4. Spencer Weart, *The Discovery of Global Warming*, new ed. (Cambridge, MA: Harvard University Press, 2008), and its constantly updated hypertext version at http://www.aip.org/history/climate/index.html (accessed February 17, 2018).

22. Guy Stewart Callendar, "The Artificial Production of Carbon Dioxide and Its Influence on Temperature," *Quarterly Journal of the Royal Meteorological Society* 64 (1938): 223–40. Callendar wrote numerous articles through the early 1960s, commenting on different aspects of climate change and its causes. See also Callendar, "Can CO_2 Influence Climate?," *Weather* 4 (1949): 310–14.

23. James R. Fleming, *The Callendar Effect: The Life and Work of Guy Stewart Callendar (1898–1964)* (Boston: American Meteorological Society, 2007); Sverker Sörlin, "Narratives and Counter Narratives of Climate Change: North Atlantic Glaciology and Meteorology, ca. 1930–1955," *Journal of Historical Geography* 35, no. 2 (2009): 237–55.

24. Sverker Sörlin, *Carl-Gustaf Rossby, 1898–1957* (Stockholm: Royal Swedish Academy of Engineering Sciences, 2015); Fleming, *Inventing Atmospheric Science*.

25. Stig Fonselius and Folke Koroleff, with a preface by Kurt Buch, "Microdetermination of CO_2 in the Air, with Current Data for Scandinavia," *Tellus* 7, no. 2 (1955): 258–65. Maria Bohn, "Concentrating on CO_2: The Scandinavian and Arctic Measurements," *Osiris* 26 (2011): 165–79.

26. In 1953, media coverage of climate change caused by carbon dioxide appeared in *Time, New York Post,* and the *Dagens Nyheter* of Stockholm, to mention just a few. Interestingly, the media interest shifted quickly and seamlessly from a nonanthropogenic interest in climate change to the modern orthodoxy. Media did not seem to bother initially with the fundamental difference in outlook, likely because the paradigmatic "environmental" understanding was not yet established.

27. "Man's Milieu," *Time* 68 (December 17, 1956): 68–79.

28. Gilbert N. Plass, "The Carbon Dioxide Theory of Climatic Change," *Tellus* 8 (1956): 140–54; Roger Revelle and Hans E. Suess, "Carbon Dioxide Exchange between Atmosphere and Ocean and the Question of an Increase of Atmospheric CO_2 during the Past Decades," *Tellus* 9 (1957): 18–27, argued along similar lines, reinforcing the revival of Callendar's hypothesis.

29. Questioning, for example, Revelle and Suess, "Carbon Dioxide Exchange." Accepting, for example, Bert Bolin and Erik Eriksson, "Distribution

of Matter in the Sea and in the Atmosphere: Changes in the Carbon Dioxide Content of the Atmosphere and Sea Due to Fossil Fuel Combustion," in *The Atmosphere and the Sea in Motion: Scientific Contributions to the Rossby Memorial Volume*, ed. Bert Bolin (New York: Rockefeller University Press, 1958), 130–42.

30. Milutin Milanković, *Théorie mathématique des phénomènes thermiques produits par la radiation solaire* (Paris: Gauthier-Villars, 1920).

31. Fred Hoyle, "External Sources of Climatic Variation," *Quarterly Journal of the Royal Meteorological Society* 75 (1949): 163.

32. Ronald E. Doel, Robert Marc Friedman, Julia Lajus, Sverker Sörlin, and Urban Wråkberg, "Strategic Arctic Science: National Interests in Building Natural Knowledge—Interwar Era through the Cold War," *Journal of Historical Geography* 42 (2014): 60–80.

33. Echoes of this appear in ecologist Charles Elton's radio broadcasts. C. S. Elton, *The Ecology of Invasions by Animals and Plants* (London: Methuen, 1958), discussed in chapter 4.

34. John von Neumann, "Can We Survive Technology?," *Fortune* (June 1955), reprinted in von Neumann, *Collected Works*, vol. 6, *Theory of Games, Astrophysics, Hydrodynamics and Meteorology* (New York: Macmillan, 1963), 513–14.

35. Jacob Darwin Hamblin, *Arming Mother Nature: The Birth of Catastrophic Environmentalism* (New York: Oxford University Press, 2013), 138.

36. Ibid., 158–62.

37. On this development, see especially Ronald Doel, "Constituting the Postwar Earth Sciences: The Military's Influence on the Environmental Sciences in the USA after 1945," *Social Studies of Science* 33, no. 5 (2003): 635–66; Alan A. Needell, *Science, Cold War and the American State: Lloyd V. Berkner and the Balance of Professional Ideals* (Amsterdam: Harwood Academic, 2000).

38. Neumann, "Can We Survive Technology?," 504.

39. Lewis Mumford, "History: Neglected Clue to Technology Change," *Technology and Culture* 2, no. 3 (1961): 230–36.

40. Lynn White Jr., "The Historical Roots of Our Ecologic Crisis," *Science* 155, no. 3767 (1967): 1203–7.

41. Terence Armstrong, *The Russians in the Arctic: Aspects of Soviet Exploration and Exploitation of the Far North, 1937–57* (London: Methuen, 1958), esp. 67–79. Matthew Farish, "Creating Cold War Climates: The Laboratories of American Globalism," in *Environmental Histories of the Cold War*, ed. J. R. McNeill and Corinna R. Unger (New York: Cambridge University Press, 2010), 51–83. Julia Lajus and Sverker Sörlin, "An Ice Free Arctic Sea? The Science of Sea Ice and Its Interests," in *Media and Arctic Climate Politics:*

Breaking the Ice, ed. M. Christensen, A. Nilsson, and N. Wormbs (New York: Palgrave Macmillan, 2013), 70–92.

42. Matthias Heymann, Henry Knudsen, Maiken L. Lolck, Henry Nielsen, and Christopher Jacob Ries, "Exploring Greenland: Science and Technology in Cold War Settings," *Scientia Canadensis* 33, no. 2 (2010): 11–42.

43. Doel et al., "Strategic Arctic Science."

44. Julia Lajus and Sverker Sörlin, "Melting the Glacial Curtain: The Politics of Scandinavian-Soviet Networks in the Geophysical Field Sciences between Two Polar Years, 1932/33–1957/58," *Journal of Historical Geography* 42 (2014): 44–59. G. Rowley, *Cold Comfort: My Love Affair with the Arctic* (1996; new ed., Montreal: McGill-Queen's Press, 2006).

45. Janet Martin-Nielsen, " 'The Deepest and Most Rewarding Hole Ever Drilled': Ice Cores and the Cold War in Greenland," *Annals of Science* 70 (2013): 1. Griffiths, "Commentary."

46. Eduard Suess, *Die Enstehung der Alpen* (The Origin of the Alps) (Vienna: W. Braunmüller, 1875). Suess also first launched the concept "biosphere," although his vegetational definition was very different from the integrative biogeochemical approach taken by Vernadsky.

47. Antoni Boleslaw Dobrowolski, *Historia naturalna lodu* (The Natural History of Ice) (Warsaw: Kasa Pomocy im. Dr. J. Mianowskiego, 1923), with a French summary. R. G. Barry, J. Jania, and K. Birkenmajer, "A. B. Dobrowolski—the First Cryospheric Scientist—and the Subsequent Development of Cryospheric Science," *History of Geo- and Space Sciences* 2, no. 1 (2011): 75–79.

48. Clarence J. Glacken, "Changing Ideas of the Habitable World," in *Man's Role in Changing the Face of the Earth*, ed. William L. Thomas Jr. (Chicago: University of Chicago Press, 1956), 86. Clarence J. Glacken, *Traces on the Rhodian Shore: Nature and Culture in Western Thought from Ancient Times to the End of the Eighteenth Century* (Berkeley: University of California Press, 1967); Grinevald, "Introduction."

49. C. W. Thornthwaite, "Modification of Rural Microclimates," in Thomas, *Man's Role in Changing*, 570. Doubravka Olsáková, ed., *In the Name of the Great Work: Stalin's Plan for the Transformation of Nature and Its Impact in Eastern Europe* (New York: Berghahn Books, 2016), esp. chap. 1 by Paul Josephson, "Introduction: The Stalin Plan for the Transformation of Nature, and the East European Experience," 1–41.

50. Sörlin, "Narratives and Counter Narratives."

51. James R. Fleming, *Fixing the Sky: The Checkered History of Weather and Climate Control* (New York: Columbia University Press, 2010); Kristine C. Harper, *Make It Rain: State Control of the Atmosphere in Twentieth-Century America* (Chicago: University of Chicago Press, 2017).

52. Study of Critical Environmental Problems (SCEP), *Man's Impact on the Global Environment: Assessment and Recommendation for Action* (Cambridge, MA: MIT Press, 1970), 18.

53. Weart, *Discovery of Global Warming*; Naomi Oreskes, "Beyond the Ivory Tower: The Scientific Consensus on Climate Change," *Science* 306, no. 5702 (2004): 1686.

54. Joshua P. Howe, *Behind the Curve: Science and the Politics of Global Warming* (Seattle: University of Washington Press, 2014).

55. Charles D. Keeling, "Is Carbon Dioxide from Fossil Fuels Changing Man's Environment?," *Proceedings of the American Philosophical Society* 114 (1970): 10–14. Kenneth Boulding, *The Meaning of the Twentieth Century: The Great Transition* (New York: Harper and Row, 1964). Harrison Brown, *The Challenge of Man's Future* (New York: Viking, 1954).

56. Lars-Göran Engfeldt, *From Stockholm to Johannesburg and Beyond: The Evolution of the International System for Sustainable Development Governance and Its Implications* (Stockholm: Ministry of Foreign Affairs, 2009).

57. C. L. Wilson and W. H. Matthews, eds., *Inadvertent Climate Modification: Report of Conference, Study of Man's Impact on Climate (SMIC), Stockholm* (Cambridge, MA: MIT Press, 1971).

58. See chapter 6. Bert Bolin, *A History of the Science and Politics of Climate Change: The Role of the Intergovernmental Panel on Climate Change* (Cambridge: Cambridge University Press, 2007), 40.

59. Ibid.

60. *Carbon Dioxide and Climate: A Scientific Assessment* (Washington, DC: National Academy of Sciences, 1979).

61. Maarten Hajer and Wytske Versteeg, "A Decade of Discourse Analysis of Environmental Politics: Achievements, Challenges, Perspectives," *Journal of Environmental Policy and Planning* 7, no. 3 (2005): 175–84.

62. Edwards, *Vast Machine*.

63. Bolin, *History of Science and Politics*.

64. For this, see also John McCormick, *The Global Environmental Movement* (London: Belhaven, 1989).

65. Rob Nixon, *Slow Violence and the Environmentalism of the Poor* (Cambridge, MA: Harvard University Press, 2011).

Chapter 6. "The Earth Is One but the World Is Not"

1. World Commission on Environment and Development, *Our Common Future* (New York: Oxford University Press, 1987), 5.

2. Ibid., 16.

3. Ibid., 5; see also 12.

4. Lynton K. Caldwell, "Environment: A New Focus for Public Policy?,"

Public Administration Review 23, no. 3 (September 1963): 132–39. In similar fashion, *Our Common Future* spoke of having to deal with "interlocking crises" (13).

5. *Our Common Future,* 28.

6. Jane Carruthers, "Tracking in Game Trails: Looking Afresh at the Politics of Environmental History in South Africa," *Environmental History* 11, no. 4 (2006): 820; John M. MacKenzie, *The Empire of Nature: Hunting, Conservation and British Imperialism* (Manchester: Manchester University Press, 1988).

7. ICSU was founded to promote international scientific activity in the different branches of science and its application for the benefit of humanity. It was built from two earlier scientific bodies, the International Association of Academies (IAA, 1899–1914) and the International Research Council (IRC, 1919–31), which hosted international scientific meetings and research. See International Council for Science, "About Us: A Brief History," https://www.icsu.org/about-us/a-brief-history (accessed March 4, 2018).

8. Rebecca Wright, Frank Trentmann, and Hiroki Shin, *From World Power Conference to World Energy Council: 90 Years of Energy Co-operation, 1923–2013* (London: World Energy Council, 2013).

9. Joseph Morgan Hodge, *Triumph of the Expert: Agrarian Doctrines of Development and the Legacies of British Colonialism* (Athens: Ohio University Press, 2007).

10. Jessica Reinisch, "Internationalism in Relief: The Birth (and Death) of UNRRA," *Past and Present,* supplement 6 (2013): 258–89

11. Cited in John McCormick, *The Global Environmental Movement* (London: Belhaven, 1989), 25.

12. See the following essays in *Nature,* no. 3967 (November 10, 1945): J. Huxley, "Science and the United Nations," 553–56; J. G. Crowther, "World Co-operation in Science," 556–57; J. D. Bernal, "A Permanent International Scientific Constitution," 557–58; J. Needham, "The Place of Science and International Scientific Co-operation in Post-war World Organization," 558–61. See also Libby Robin and Will Steffen, "History for the Anthropocene," *History Compass* (2007): 1694–1719, esp. 1699–1700.

13. C. P. Snow, *The Two Cultures and the Scientific Revolution* (Cambridge: Cambridge University Press, 1959), 6.

14. Jon Agar, *Science in the Twentieth Century and Beyond* (Cambridge: Polity Press, 2012), 264–66, 306–8; David Edgerton, *Britain's War Machine: Weapons, Resources and Experts in the Second World War* (London: Allen Lane, 2011).

15. Joseph Morgan Hodge, *Triumph of the Expert: Agrarian Doctrines of Development and the Legacies of British Colonialism* (Athens: University of Ohio Press, 2007).

16. Edmund Russell, *War and Nature: Fighting Humans and Insects with Chemicals from World War I to "Silent Spring"* (Cambridge: Cambridge University Press, 2001).

17. Agar, *Science in the Twentieth Century*, 291–94, 302–8, 373–85.

18. Paul N. Edwards, *A Vast Machine: Computer Models, Climate Data, and the Politics of Global Warming* (Cambridge, MA: MIT Press, 2010).

19. Eglė Rindzevičiūtė, *The Power of Systems: How Policy Sciences Opened Up the Cold War World* (Ithaca, NY: Cornell University Press, 2016). Hoos is cited on page 75.

20. Alison Bashford, *Global Population: History, Geopolitics and Life on Earth* (New York: Columbia University Press, 2014), 206–8, 270–78.

21. Patrick Petitjean, V. Zharov, G. Glaser, J. Richardson, B. de Paderac, and G. Archibald, eds., *Sixty Years of Science at UNESCO, 1945–2005* (Paris: UNESCO Publishing, 2006), 21; Stanley Johnson, *UNEP: The First Forty Years* (Nairobi: United Nations Environment Programme, 2013), https://europa.eu /capacity4dev/unep/document/book-unep-first-forty-years-stanley-johnson.

22. Robin and Steffen, "History for the Anthropocene"; Gail Archibald, "How the 'S' Came to Be in UNESCO," in Petitjean et al., *Sixty Years of Science*; V. Enebakk, "UNESCO and the History of Science and Its Social Relations" (paper presented at British Society for the History of Science conference, "Science and Its Social Relations," London, September 17, 2006); paper kindly provided by the author. The League of Nations established an International Institute for Intellectual Co-operation (IIIC) in 1928.

23. Bashford, *Global Population*.

24. Paul Robbins and Sarah A. Moore, "Ecological Anxiety Disorder: Diagnosing the Politics of the Anthropocene," *Cultural Geographies* 20, no. 1 (2012).

25. McCormick, *Global Environmental Movement*, 34–35.

26. Stephen J. Macekura, *Of Limits and Growth: The Rise of Global Sustainable Development in the Twentieth Century* (New York: Cambridge University Press, 2015).

27. McCormick, *Global Environmental Movement*, 46.

28. Macekura, *Of Limits and Growth*, 61–63.

29. Roderick Nash, *Wilderness and the American Mind* (1967; rev. 5th ed., New Haven, CT: Yale University Press, 2014).

30. Aldo Leopold, *A Sand County Almanac and Sketches Here and There* (1949; repr., Oxford: Oxford University Press, 1987); Donald Worster, *A River Running West: The Life of John Wesley Powell* (Oxford: Oxford University Press, 2000).

31. Nash, *Wilderness and the American Mind*; Libby Robin, "Being First: Why the Americans Needed It, and Why Royal National Park Didn't Stand

in Their Way," *Australian Zoologist* (2013): 321–29; B. Adams, "Once the Wild Is Gone," *Aeon*, October 23, 2012, https://aeon.co/essays/the-wilderness-fetish-is-bad-for-people-and-for-the-planet.

32. Mark Wilson, "The British Environmental Movement: The Development of an Environmental Consciousness and Environmental Activism, 1945–1975" (PhD thesis, University of Northumbria, Newcastle, 2014).

33. IUCN (Fontainbleau), "Article 1, Objects" (Brussels: IUCN Library, 1948), 17, http://data.iucn.org/dbtw-wpd/edocs/1948-001.pdf.

34. Libby Robin, "Nature Conservation as a National Concern," *Historical Records of Australian Science* 10, no. 1 (1994): 1–24.

35. James C. Scott, *Seeing Like a State* (New Haven, CT: Yale University Press, 1998); Theodore Porter, *Trust in Numbers* (Princeton, NJ: Princeton University Press, 1996).

36. Peter Crowcroft, *Elton's Ecologists: A History of the Bureau of Animal Population* (Chicago: University of Chicago Press, 1991).

37. *National Parks Review: A Discussion Document* (Cheltenham: Countryside Commission, 1989); *National Parks in England and Wales*, Parliamentary Papers 1944/5, Cmd 6628, v, 283–339 (Dower Report). There were no national parks in Scotland until the twenty-first century.

38. For a selection of the very large literature dealing with this theme, see Bernhard Gissibl, Sabine Höhler, and Patrick Kupper, eds., *Civilizing Nature: National Parks in a Global Historical Perspective* (New York: Berghan, 2012); Patrick Kupper, *Wildnis schaffen: Eine transnational Geschichte des Schweizerischen Nationalparks* (Bern: Haupt Verlag, 2012); Richard West Sellars, *Preserving Nature in the National Parks: A History* (New Haven, CT: Yale University Press, 1997); Sverker Sörlin, "Monument and Memory: Landscape Imagery and the Articulation of Territory," *Worldviews: Environment, Culture, Religion* 2 (1998): 269–79.

39. Wilderness and natural heritage management was important at this time in other places, such as Australia. Robin, "Being First."

40. Macekura, *Of Limits and Growth*, 242.

41. Marcus Haward and Tom Griffiths, eds., *Australia and the Antarctic Treaty System: 50 Years of Influence* (Sydney: University of New South Wales Press, 2011). Note: Oxford dictionary gives 1962 as a starting year for the use of the term *environmental science*, but usage in the American military of the concept was earlier. Ronald E. Doel, "Constituting the Postwar Earth Sciences: The Military's Influence on the Environmental Sciences in the USA after 1945," *Social Studies of Science* 33, no. 50 (2003): 635–66.

42. The FAO took over some of the functions of the International Institute for Agriculture, founded in Rome in 1905.

43. Iain McCalman, *The Reef: A Passionate History* (Melbourne: Penguin,

2014); Ken Conca, *An Unfinished Foundation: The United Nations and Global Environmental Governance* (Oxford: Oxford University Press, 2015), 45.

44. Edwards, *A Vast Machine*.

45. WMO originated from the International Meteorological Organization (IMO), which was founded in 1873.

46. Johnson, *UNEP*, 17; UNESCO, "Man and the Biosphere Programme," http://www.unesco.org/new/en/natural-sciences/environment/ecological -sciences/man-and-biosphere-programme (accessed August 4, 2013).

47. Spencer R. Weart, *The Discovery of Global Warming* (2003; rev. and expanded ed., Cambridge, MA: Harvard University Press, 2008); Edwards, *A Vast Machine*, 361–62.

48. Edwards, *A Vast Machine*, 361–72.

49. The United Nations Environmental Programme website is www .unenvironment.org. The scientific advisory groups are the Ecosystem Conservation Group (ECG), the Intergovernmental Panel on Climate Change (IPCC), the Joint Group of Experts on the Scientific Aspects of Marine Environmental Protection (GESAMP), the Scientific and Technical Advisory Panel (STAP), and the UN Scientific Committee on the Effects of Atomic Radiation (UNSCEAR).

50. Future Earth: Research for Global Sustainability, http://www.future earth.org/about. The projects—most of which have acronyms—are listed at http://www.futureearth.org/research (including reference to the earlier forms, supported by ICSU) (accessed March 26, 2018).

51. J. Rockström et al., "A Safe Operating Space for Humanity," *Nature* 461 (2009): 472–75.

52. Spencer Weart, "The Development of the Concept of Dangerous Anthropogenic Climate Change," in *The Oxford Handbook of Climate Change and Society*, ed. John S. Dryzek, Richard B. Norgaard, and David Schlosberg (Oxford: Oxford University Press, 2011). The official title of the conference was the World Conference on the Changing Atmosphere.

53. Mike Hulme, *Exploring Climate Change through Science and in Society: An Anthology of Mike Hulme's Essays, Interviews and Speeches* (London: Routledge, 2013); Intergovernmental Panel on Climate Change, *Climate Change 2007: The Physical Science Basis*, contribution of Working Group 1 to the Fourth Assessment Report of the IPCC, ed. S. Solomon et al. (Cambridge: Cambridge University Press, 2007).

54. UN Environment, http://www.unenvironment.org.

55. The American Presidency Project, "Lyndon B. Johnson," http://www .presidency.ucsb.edu/ws/?pid=27355 (accessed October 7, 2015).

56. Thomas Robertson, *The Malthusian Moment: Global Population Growth*

and the Birth of American Environmentalism (New York: Rutgers University Press, 2012), 166–68.

57. Jens Ivo Engels, *Naturpolitik in der Bundesrepublik: Ideenwelt und politische Verhaltensstile in Naturschutz und Umweltbewegung 1950–1980* (Paderborn: Ferdinand Schöningh, 2006), 275. See also the discussion in Holger Nehring, "Genealogies of the Ecological Moment: Planning, Complexity and the Emergence of 'the Environment' as Politics in West Germany, 1949–1982," in Sverker Sörlin and Paul Warde, eds., *Nature's End: History and the Environment* (Basingstoke: Palgrave Macmillan, 2009), 115–38.

58. Sandra Chaney, *The Nature of the Miracle Years: Conservation in West Germany, 1945–1975* (New York: Berghahn, 2008), 176–94; Ute Hasenöhrl, *Zivilgesellschaft und Protest: Eine geschichte der Naturschutz und Umweltbewegung in Bayern, 1945–1980* (Göttingen: Vandenhoeck and Ruprecht, 2008), 266–69.

59. Hasenöhrl, *Zivilgesellschaft und Protest*, 271.

60. John Sheail, *An Environmental History of Twentieth-Century Britain* (Basingstoke: Palgrave, 2002), 271–72. See also Susan Owens, *Knowledge, Policy and Expertise: The UK Royal Commission on Environmental Pollution, 1970–2011* (Oxford: Oxford University Press, 2015); Florian Charvolin, "L'année clef pour la définition de l'environnement en France," *La Revue pour l'Histoire du CNRS* 4 (2001).

61. McCormick, *Global Environmental Movement*, 125.

62. Sunayana Ganguly, *Deliberating Environmental Policy in India: Participation and the Role of Advocacy* (Oxford: Routledge, 2016), 7–12.

63. Charles McElwee, *Environmental Law in China: Mitigating Risk and Ensuring Compliance* (Oxford: Oxford University Press, 2011); Rachel E. Stern, *Environmental Litigation in China: A Study in Political Ambivalence* (Cambridge: Cambridge University Press, 2013), 37–42; Lee Liu, "Sustainability Efforts in China: Reflections on the Environmental Kuznets Curve through a Local Evaluation of 'Eco-Communities,'" *Annals of the Association of American Geographers* 98, no. 3 (2008).

64. See, for example, Jairam Ramesh, *Green Signals: Ecology, Growth, and Democracy in India* (Oxford: Oxford University Press, 2015); Arun Agrawal, *Environmentality: Technologies of Government and the Making of Subjects* (Durham, NC: Duke University Press, 2005).

65. See, for example, Thomas Le Roux, *Le laboratoire des pollutions industrielles: Paris, 1770–1830* (Paris: Albin Michel, 2011); Stephen Mosley, *The Chimney of the World: A History of Smoke Pollution in Victorian and Edwardian Manchester* (Cambridge: White Horse Press, 2001); Franz-Josef Brüggemeier and Thomas Rommelspacher, *Blauer Himmel über der Ruhr: Geschichte*

der Umwelt im Ruhrgebiet, 1840–1990 (Essen: Klartext, 1992); Christoph Bern-hardt and Geneviève Massard-Guilbaud, *Le demon modern: La pollution dans les sociétés urbaines et industrielles d'Europe* (Clermont-Ferrand: Presses Uni-versitaires Blaise-Pascal, 2002).

66. See Chaney, *Nature of the Miracle Years*, 162; also, Britain's Conser-vation Society was most active between 1966 and 1973 and acted as a transi-tional organization uniting figures from previous conservation movements and the organic farming campaigners of the Soil Association with Mal-thusian population theories. Horace Herring, "The Conservation Society: Harbinger of the 1970s Environment Movement in the UK," *Environment and History* 7 (2001): 381–401.

67. Samuel P. Hays, *Conservation and the Gospel of Efficiency: The Progres-sive Conservation Movement, 1890–1920* (Pittsburgh: University of Pittsburgh Press, 1959).

68. Paige West, *Conservation Is Our Government Now: The Politics of Ecol-ogy in Papua New Guinea* (Durham, NC: Duke University Press, 2006).

69. Ramachandra Guha, *Environmentalism: A Global History* (New York: Longman, 2000), 98–124.

70. Macekura, *Of Limits and Growth*, 113–15.

71. Conca, *Unfinished Foundation*, 38–39.

72. McCormick, *Global Environmental Movement*, 106–24.

73. Macekura, *Of Limits and Growth*, 196–218, 261–99.

74. Arun Agrawal, *Environmentality: Technologies of Government and the Making of Subjects* (Durham, NC: Duke University Press, 2005); Richard A. Walker, *The Country in the City: The Greening of the San Francisco Bay Area* (Seattle: University of Washington Press, 2007).

75. Xuemei Bai, Richard J. Dawson, Diana Ürge-Vorsatz, Gian C. Delgado, Aliyu Salisu Barau, Shobhakar Dhakal, David Dodman, Lykke Leonardsen, Valérie Masson-Delmotte, Debra C. Roberts, and Seth Schultz, "Six Research Priorities for Cities and Climate Change," *Nature* 555 (2018): 23–25.

76. Thomas Elmqvist, Xuemei Bai, Niki Frantzeskaki, Corrie Griffith, David Maddox, Timon McPhearson, Susan Parnell, Debra Roberts, Patricia Romero-Lankao, David Simon, and Mark Watkins, eds., *Urban Planet* (Cambridge: Cambridge University Press, 2018).

77. Frank Zelko, *Make It a Green Peace! The Rise of a Countercultural Envi-ronmentalism* (New York: Oxford University Press, 2013).

78. Rachel Rothschild, "Acid Wash: How Cold War Politics Helped Solve a Climate Crisis," *Foreign Affairs*, August 24, 2015, https://www.foreignaffairs.com/articles/2015–08–24/acid-wash; Conca, *Unfinished Foundation*, 43, 73; McCormick, *Global Environmental Movement*, 174, 182–87; Rindzevičiūtė, *Power of Systems*, 180–203.

79. Charles Taylor, *Modern Social Imaginaries* (Durham, NC: Duke University Press, 2004), 23.

80. Timothy Sinclair, *Global Governance* (Cambridge: Polity Press, 2012), 20.

Chapter 7. Seeking a Safe Future

1. J. Rockström, W. Steffen, K. Noone, Å. Persson, F. S. Chapin III, E. Lambin, T. M. Lenton, M. Scheffer, C. Folke, H. Schellnhuber, B. Nykvist, C. A. De Wit, T. Hughes, S. van der Leeuw, H. Rodhe, S. Sörlin, P. K. Snyder, R. Costanza, U. Svedin, M. Falkenmark, L. Karlberg, R. W. Corell, V. J. Fabry, J. Hansen, B. Walker, D. Liverman, K. Richardson, P. Crutzen, and J. Foley, "Planetary Boundaries: Exploring the Safe Operating Space for Humanity," *Ecology and Society* 14, no. 2 (2009), http://www.ecologyandsociety.org/vol 14/iss2/art32.

2. Gretchen Daily, *Nature's Services: Societal Dependence on Natural Ecosystems* (Washington, DC: Island Press, 1997). For a critical survey, see Richard B. Norgaard, "Ecosystem Services: From Eye-Opening Metaphor to Complexity Blinder," *Ecological Economics* 69 (2010): 1219–27.

3. J. Rockström et al., "A Safe Operating Space for Humanity," *Nature* 461 (2009): 472–75.

4. The idea of resilience had been particularly promoted by Canadian ecologist Buzz Holling after a seminal article in 1973. Holling went on to play a major role in approaches to system modeling at IIASA (see chapter 6). Eglė Rindzevičiūtė, *The Power of Systems: How Policy Sciences Opened Up the Cold War World* (Ithaca, NY: Cornell University Press, 2016), 101; C. S. Holling, "Resilience and Stability of Ecological Systems," *Annual Review of Ecology and Systematics* 4 (1973): 1–23.

5. It is still evolving as this book goes to press. See http://futureearth.org /projects.

6. Rindzevičiūtė, *Power of Systems*, 141–42, 176–79.

7. Ariane Tanner, *Die Mathematisierung des Lebens: Alfred Lotka und der energetische Holismus im 20. Jahrhundert* (Tübingen: Mohr Siebeck, 2017), 119.

8. John Sheail, *The Natural Environment Research Council—A History* (Swindon: Natural Environment Research Council, 1992).

9. James Lovelock and Lynn Margulis, "Atmospheric Homeostasis by and for the Biosphere: The Gaia Hypothesis," *Tellus* 26 (1974): 2.

10. S. R. Weart, *The Discovery of Global Warming* (Cambridge, MA: Harvard University Press, 2003), 150. T. F. Malone, preface to *Global Change: The Proceedings of a Symposium Sponsored by the International Council of Scientific Unions (ICSU) during Its 20th General Assembly in Ottawa, Canada on*

September 25, 1984, ed. T. F. Malone and J. G. Roederer (Cambridge: ICSU Press, 1985), xviii. Chunglin Kwa, "Speaking to Science: The Programming of Interdisciplinary Research through Informal Science-Policy Interactions," *Science and Public Policy* 33, no. 6 (2006): 457–67.

11. Ola Uhrqvist & Björn-Ola Linnér, "Narratives of the Past for Future Earth: The Historiography of Global Environmental Change Research," *Anthropocene Review* 2 (2015): 159–73.

12. Robert Costanza, Lisa J. Graumlich, and Will Steffen, eds., *Sustainability or Collapse? An Integrated History and Future of People on Earth* (Cambridge, MA: MIT Press, 2006); W. Steffen, A. Sanderson, P. D. Tyson, J. Jäger, P. A. Matson, B. Moore III, F. Oldfield, K. Richardson, H. J. Schellnhuber, B. L. Turner, and R. J. Wasson, "Executive Summary" of *Global Change and the Earth System: A Planet under Pressure* (Berlin: Springer-Verlag, 2004), 4–39. John McNeill and Peter Engelke, *The Great Acceleration: An Environmental History of the Anthropocene since 1945* (Cambridge, MA: Harvard University Press, 2016). On the genealogy of the Great Acceleration concept, see Will Steffen, Wendy Broadgate, Lisa Deutsch, Owen Gaffney, and Cornelia Ludwig, "The Trajectory of the Anthropocene: The Great Acceleration," *Anthropocene Review* 2 (2015): 81–98.

13. Michel Foucault, "Governmentality," trans. Rosi Braidotti and revised by Colin Gordon, in *The Foucault Effect: Studies in Governmentality*, ed. Graham Burchell, Colin Gordon, and Peter Miller, 87–104 (Chicago: University of Chicago Press, 1991). See also Frank Fischer, *Democracy and Expertise: Reorientating Policy Enquiry* (Oxford: Oxford University Press, 2009).

14. There is a comprehensive literature on governmentality and its application to environmental issues. See, for example, Bruce Braun, "Producing Vertical Territory: Geology and Governmentality in Late Victorian Canada," *Ecumene* 7, no. 1 (2000): 7–46; A. Agrawal, *Environmentality: Technologies of Government and the Making of Subjects* (Durham, NC: Duke University Press, 2005); Kristin Asdal, "The Problematic Nature of Nature: The Post-Constructivist Challenge to Environmental History," *History and Theory* 42, no. 4 (2003): 239–53; Ola Uhrqvist, *Seeing and Knowing the Earth as a System: An Effective History of Global Environmental Change Research as Scientific and Political Practice* (PhD diss., Linköping University, 2014). See also Gavin Bridge and T. Perreault, "Environmental Governance," in *A Companion to Environmental Geography*, ed. Noel Castree, 475–97 (Chichester: Wiley-Blackwell, 2009). On the growth of ecosystem services, see Henrik Ernstson and Sverker Sörlin, "Ecosystem Services as Technology of Globalization: On Articulating Values in Urban Nature," *Ecological Economics* 86 (2013): 273–84.

15. J.-B. Fressoz and C. Bonneuil, *The Shock of the Anthropocene: The Earth, History, and Us* (London: Verso, 2015), 88.

16. Rob Nixon, *Slow Violence and the Environmentalism of the Poor* (Cambridge, MA: Harvard University Press, 2011). Frantz Fanon, *Les damnés de la terre* (Paris: Éditions Maspero, 1961), translated into English as *The Wretched of the Earth* (1965).

17. All now part of Future Earth. See http://www.futureearth.org/projects /ihope-integrated-history-and-future-people-earth (accessed 4 March 2018).

18. David Christian, *Maps of Time: An Introduction to Big History* (Berkeley: University of California Press, 2004); Ian Morris, *Why the West Rules— for Now: The Patterns of History and What They Reveal about the Future* (London: Profile Books, 2010); John L. Brooke, *Climate Change and the Course of Global History: A Rough Journey* (Cambridge: Cambridge University Press, 2014).

19. Helge Jordheim, "Introduction: Multiple Times and the Work of Synchronization," *History and Theory* 53 (2014): 498–518.

20. Will Steffen, "Commentary," in *The Future of Nature: Documents of Global Change*, ed. Libby Robin, Sverker Sörlin, and Paul Warde, 486–90 (New Haven, CT: Yale University Press, 2013).

21. P. J. Crutzen, "Geology of Mankind," *Nature*, no. 415 (2002): 23; P. J. Crutzen and E. F. Stoermer, "The 'Anthropocene,'" *IGBP Newsletter*, no. 41 (2000): 17–18.

22. Robert Macfarlane, "Generation Anthropocene," *Guardian*, April 1, 2016, http://www.theguardian.com/books/2016/apr/01/generation-anthropo cene-altered-planet-for-ever. See also "Welcome to the Anthropocene" and "Briefing the Anthropocene: A Man-Made World," *Economist*, May 28, 2011.

23. Vicky Albritton and Fredrik Albritton Jonsson, *Green Victorians: The Simple Life in John Ruskin's Lake District* (Chicago: University of Chicago Press, 2016). Eugène Huzar, *La fin du monde par la science* (Paris: Dentu, 1855).

24. S. C. Finney and L. E. Edwards, "The Anthropocene Epoch: Scientific Decision or Political Statement?' *GSA Today* 26, no. 3 (2016): 4–10.

25. Paul Warde, Libby Robin, and Sverker Sörlin, "Stratigraphy for the Renaissance: Questions of Expertise for 'the Environment' and 'the Anthropocene,'" *Anthropocene Review* 4, no. 3 (2017): 246–58.

26. Eileen Crist, "On the Poverty of Our Nomenclature," *Environmental Humanities* (2013). Donna Haraway, "Anthropocene, Capitalocene, Plantationocene, Chthulucene: Making Kin," *Environmental Humanities* (2015). Andreas Malm and Alf Hornborg, "The Geology of Mankind? A Critique of the Anthropocene Narrative," *Anthropocene Review* (2014). Gísli Pálsson, Bronislaw Szerzynski, Sverker Sörlin et al., "Reconceptualizing the 'Anthro-

pos' in the Anthropocene: Integrating the Social Sciences and Humanities in Global Environmental Change Research," *Environmental Science and Policy* 28 (2013): 3–13.

27. Marco Armiero and Massimo De Angelis, "Anthropocene: Victims, Narrators, and Revolutionaries," *South Atlantic Quarterly* 116, no. 2 (April 2017): 353.

28. Ghassan Hage, *Is Racism an Environmental Threat?* (Cambridge: Polity, 2017); Benjamin Hale, *The Wild and the Wicked: On Nature and Human Nature* (Cambridge, MA: MIT Press, 2016).

29. Fischer, *Democracy and Expertise*; Jedediah Purdy, *After Nature: A Politics for the Anthropocene* (Cambridge, MA: Harvard University Press, 2015), calls for the articulation of a new responsible "earth politics." See Sverker Sörlin, *Anthropocene—An Essay on the Age of Humanity* (in Swedish) (Stockholm: Weyler, 2017), chap. 11. See also Simon Nicholson, ed., *New Earth Politics: Essays from the Anthropocene* (Cambridge, MA: MIT Press, 2016).

30. Mike Hulme, "Reducing the Future to Climate: A Story of Climate Determinism and Reductionism," *Osiris* 26 (2011): 245–66.

31. Arjun Appadurai, *The Future as Cultural Fact: Essays on the Global Condition* (London: Verso, 2013), 285–87.

32. Dipesh Chakrabarty, "Climate and Capital: On Conjoined Histories," *Critical Inquiry* 41, no. 1 (Autumn 2014): 1–23.

33. Libby Robin and Cameron Muir, "Slamming the Anthropocene: Performing Climate Change in Museums," *reCollections* 10, no. 1 (2015), http://recollections.nma.gov.au/issues/volume_10_number_1/papers/slamming_the_anthropocene.

34. Katherine Gibson, Deborah Bird Rose, and Ruth Fincher, eds., *Manifesto for Living in the Anthropocene* (New York: Punctum, 2015), vii; Val Plumwood, *Environmental Culture: The Ecological Crisis of Reason* (London: Routledge, 2002), 8–9.

35. International Panel on Social Progress, "Rethinking Society for the 21st Century," in press, http://www.ipsp.org (accessed February 21, 2018).

36. Tom Griffiths, "The Humanities and an Environmentally Sustainable Australia," *Australian Humanities Review* 43 (2007), http://www.australianhumanitiesreview.org/archive/Issue-December-2007/EcoHumanities/EcoGriffiths.html; Ursula Heise, "Introduction: Planet, Species, Justice—and the Stories We Tell about Them," in *The Routledge Companion to the Environmental Humanities*, ed. Ursula Heise, Jon Christensen, and Michelle Niemann, 1–10 (London: Routledge, 2017). Ursula K. Heise, "Environmental Literature and the Ambiguities of Science," *Anglia* 133, no. 1 (2015): 22–36.

37. Amitav Ghosh, *The Great Derangement: Climate Change and the Unthinkable* (Chicago: University of Chicago Press, 2016).

38. David Lowenthal, *George Perkins Marsh: Prophet of Conservation*, foreword by William Cronon (1958; rev. ed., Seattle: University of Washington Press, 2015).

39. Mike Hulme, "Meet the Humanities," *Nature Climate Change* 1 (2011): 177–79; see also Sverker Sörlin, "Environmental Humanities: Why Should Biologists Interested in the Environment Take the Humanities Seriously?," *BioScience* 62, no. 9 (2012): 788–89; Noel Castree et al., "Changing the Intellectual Climate," *Nature Climate Change* 4, no. 9 (2014): 763–68.

40. Libby Robin, "Environmental Humanities and Climate Change: Understanding Humans Geologically and Other Life Forms Ethically," *WIREs Climate Change* 9, no. 1 (January-February 2018): 9:e499 (18 pp).

41. Appadurai, *Future as Cultural Fact*, 285–87.

42. Ibid., 287.

43. Rob Nixon, *Slow Violence and the Environmentalism of the Poor* (Cambridge, MA: Harvard University Press, 2011).

44. Alan Weisman, *The World without Us* (New York: Thomas Dunne, 2007); see also the History Channel version, *Life after People*. Elizabeth Kolbert, *The Sixth Extinction: An Unnatural History* (New York: Holt, 2014).

45. This is something Australian Aboriginal people do, as anthropologist Deborah Rose has described. Deborah Rose, "Anthropocene Noir," *Arena* 41–42 (2013–14): 206–19.

46. Naomi Klein, "Let Them Drown," reprinted in *Will the Flower Slip through the Asphalt? Writers Respond to Climate Change*, ed. Vijay Prashad, 29–47 (New Delhi: Leftword Books, 2017), 40.

47. Carolyn Merchant, *Reinventing Eden: The Fate of Nature in Western Culture* (London: Routledge, 2003).

Bibliographic Essay

The environment emerged in the postwar moment of the late 1940s and became ever more prominent with the idea of the Anthropocene, the age of humans, in the new millennium. This book follows the arc of the Great Acceleration of global environmental change that has occurred between 1945 and now. For a history of the Great Acceleration, see John McNeill and Peter Engelke, *The Great Acceleration: An Environmental History of the Anthropocene since 1945* (2016). John McNeill's magisterial *Something New under the Sun: An Environmental History of the Twentieth-Century World* (2000) also provides a history of the accelerating dependence on fossil fuels that shaped the century and its ideas about people and the environment. The associated history of environmentalism as an international political movement is surveyed in John McCormick, *The Global Environmental Movement* (1989).

The precursor to this book, *The Future of Nature: Documents of Global Change*, edited by Libby Robin, Sverker Sörlin, and Paul Warde (2013), offers a useful anthology of historical sources that have shaped and informed the idea of the environment and its key concepts. Many of the crucial sources are available in short form in this volume, including Hans Carl von Carlowitz's 1713 ideas about *Nachhaltigkeit* (sustainability), Thomas Malthus on population (1798), Wallace S. Broecker on dangerous climate change (1987), and Paul Crutzen and Eugene Stoermer on the Anthropocene (2000). In addition, the commentators in this book include authors of many important syntheses such as historian Alison Bashford, author of *Global Population* (2015), and climate scientist Mike Hulme, author of seminal books on climate change science and controversy, such as *Why*

We Disagree about Climate Change (2009) and *Weathered: Cultures of Climate* (2017).

The present book opens with a suite of postwar "jeremiads," warning of the disasters ahead if people continue to wage war on nature. Key books among these include William Vogt's *Road to Survival* (1948), Fairfield Osborn's *Our Plundered Planet* (1948), and Paul Sears's *Deserts on the March* (1935; rev. ed., 1949). The genre of "environmental alarmism" was joined soon after by nuclear scientist Harrison Brown's *The Challenge of Man's Future* (1954) and Rachel Carson's *Silent Spring* (1962). Useful commentaries on these works include Thomas Robertson, *The Malthusian Moment* (2012); Linda Lear, *Rachel Carson: Witness for Nature* (1997); Lisa H. Sideris and Kathleen D. Moore, eds., *Rachel Carson: Legacy and Challenge* (2008). Sharon E. Kingsland's *Modeling Nature: Episodes in the History of Population Ecology* (1985) provides an excellent analysis of the growth of ecological thinking in this era, its connections with demography, and the role of Alfred Lotka. Kingsland's later *The Evolution of American Ecology, 1890–2000* (2005) is one of several important histories of ecology and its role in alerting the wider public to the changing environment. Others include Donald Worster, *Nature's Economy* (1977), Joel B. Hagen, *An Entangled Bank: The Origins of Ecosystem Ecology* (1992), and Gregg Mitman, *The State of Nature* (1992). Lynn K. Nyhart's *Modern Nature* (2009) is particularly useful on the history of German ecological thinking.

After 1948, the environment became something more than "nature," and this book argues that ecology was not the only expertise needed to articulate it. Many of the ideas that attached themselves to "the environment," including questions of population, pollution, "foul" air, public health, energy and resources—and later, economics, climate, and sustainability—had important precursors in the nineteenth century and earlier. As they were not yet unified under the category of environment, unsurprisingly they have no general history, but many aspects are touched upon in David Arnold, *The Problem of Nature: Environment, Culture and European Expansion* (1996) and Paul Warde, *The Invention of Sustainability* (2018).

The suite of different sciences and of interdisciplinary expertise that engaged with the environment was already diversifying in the 1950s, exemplified in *Man's Role in Changing the Face of the Earth* (1956), edited by W. L. Thomas. Another important interdisciplinary event in the 1950s was the International Geophysical Year (IGY) 1957–58. While it did not set out to be an environmental research enterprise, this very large scientific collaboration of more than sixty nations and thousands of scientists moved geophysical and climatic issues higher on the agenda, thus facilitating their subsequent affiliation both with the idea of the environment and the environmental sciences. Charles David Keeling's measurements of carbon dioxide at the Mauna Loa Observatory in Hawaii were an IGY project from 1958. The role of the IGY is addressed in Mark Bowen's *Thin Ice: Unlocking the Secrets of Climate in the World's Highest Mountains* (2005). The history of climate science is covered by Spencer Weart, *The Discovery of Global Warming* (2003, 2008), and James R. Fleming, *Historical Perspectives on Climate Change* (1998).

Along with Rachel Carson's warnings about pesticide use and industrial agriculture, the 1960s and 1970s brought the idea of limits into focus in the emerging field of environmental studies. Notable books include R. Buckminster Fuller's *Operating Manual for Spaceship Earth* (1969), D. L. Meadows, *The Limits to Growth* (1972), and Barbara Ward and René Dubos's *Only One Earth* (1972), prepared for the UN conference on the Human Environment in Stockholm. Sabine Höhler's history, *Spaceship Earth in the Environmental Age, 1960–1990* (2015), provides an overview of limits and the role of space science in environmental thinking. The international politics of environment grew rapidly in this period as documented by Joachim Radkau in *The Age of Ecology* (2014; German ed., *Die Ära der Ëcologie*, 2011). A broad and very readable account of the debates in resource economics and in the wake of the *Limits to Growth* report can be found in Paul Sabin, *The Bet* (2013).

We have, together and as individual scholars, contributed to the expanding Anthropocene debates. Sverker Sörlin has published *Anthropocene—An Essay on the Age of Humanity* (2017, in Swedish).

New article-length histories include Paul Warde, Libby Robin, and Sverker Sörlin, "Stratigraphy for the Renaissance: Questions of Expertise for 'the Environment' and 'the Anthropocene,'" *Anthropocene Review* 4, no. 3 (2017): 246–58, and Libby Robin, "Environmental Humanities and Climate Change: Understanding Humans Geologically and Other Life Forms Ethically," *WIREs Climate Change* (2017): e499. Two brief versions of the history of the concept are Paul Warde, "The Environment," in *Local Places, Global Processes*, edited by Peter Coates, David Moon, and Paul Warde, 32–46 (Oxford: Windgather Press, 2016), and Sverker Sörlin, "Environment," in *Companion to Environmental Studies*, edited by Noel Castree, Mike Hulme, and James D. Proctor (New York: Routledge, 2018).

Our own thinking has been helpfully shaped by many encounters with books not strictly concerned with environmental history or the history of science. Among these we mention a few that have been particularly important: Theodore Porter's *Trust in Numbers: The Pursuit of Objectivity in Science and Public Life* (1995); Harry Collins and Robert Evans, *Rethinking Expertise* (2009); and Sheila Jasanoff, *States of Knowledge: The Co-production of Science and the Social Order* (2004).

Index